Aristide GANGPENDE

Interactions between flooding and water-related diseases in Chad

Aristide GANGPENDE

Interactions between flooding and water-related diseases in Chad

A study in urban areas: the case of District 7 of N'Djamena city

ScienciaScripts

Imprint

Cover image: www.ingimage.com

This book is a translation from the original published under ISBN 978-620-6-71905-2.

Publisher:
Sciencia Scripts
is a trademark of
Dodo Books Indian Ocean Ltd. and OmniScriptum S.R.L publishing group

120 High Road, East Finchley, London, N2 9ED, United Kingdom
Str. Armeneasca 28/1, office 1, Chisinau MD-2012, Republic of Moldova, Europe
Managing Directors: Ieva Konstantinova, Victoria Ursu
info@omniscriptum.com

Printed at: see last page
ISBN: 978-620-8-60516-2

Contents

A study of water-related health risks in an urban environment: the case of the 7th arrondissement of the city of N'Djaména".

district of the city of N'Djaména".

Epigraph

"We are doctors with a spatial reference, doctors of people in their living environment. in their living environment..."

Professor TCHOTSOUA Michel,

Dedication

In memory of our Dissertation Director
The late Professor TCHOTSOUA Michel.

Acknowledgements

This booklet is based on a field study carried out as part of our research dissertation in Health Geography at the University of Ngaoundéré in Cameroon.
We would like to thank all the individuals and legal entities who have contributed, one way or another, to its achievement.
We would like to express our sincere thanks to Professor Michel TCHOTSOUA (may he rest in peace) who, in spite of his many occupations and responsibilities, accepted to direct this research and put all his energy into it. If this work, in its present state, has any scientific value, it is largely thanks to his rigorous and non-paternalistic observations and intrusions.
We would like to thank all the teaching staff at the University of Ngaoundéré, particularly those in the Geography Department, for the quality of the training provided. In particular, we would like to thank Professors Wakponou Anselme, Joseph-Pierre Ndamé, Iya Moussa, Aoudou Doua, Doctors Mabouloum Anne-Marie, Anaba Christian and Kouedjou Idriss Landry.
We would like to express our deep gratitude to Mr Razak Alsalam Gdana Sali for taking time to guide and monitor this work. We would also like to thank Dr Salomon Tamira for the corrections he made to this work.
We would like to thank our big brother Djimasra Alexis (aka Mac Alex) for all the support he has given us.
Our thanks also go to the entire Gangpendé family for the many ways in which they supported my education.
Dr Diane Irie, who facilitated our integration into the various health institutions for data collection.
I would like to *thank* all the members of the *Géo 2015 class* for their fruitful exchanges, particular Kalténé Ngarhounoum Francis, Abdel Hadi, Woumou Woumbondi Gérard, Godjé Bachirou, Gwa Bedi Lucien, Dentalé Moussa and Tchoubeube Zepelin.

SUMMARY

The issue of diseases linked to rainwater (flooding) is a major concern for the 7th arrondissement of N'Djaména (Chad). In August 2020, 1,300 households were affected, with 6,553 people affected in the Amtoukoui district alone. Illnesses linked to direct or indirect contact with water account for more than 35% of the ten main causes of consultation in the 7th arrondissement. The aim of this study is to analyse the population's reactions to flooding and water-related illnesses. To this end, the analytical matrix this research was based on fieldwork, in particular data collection (questionnaires, interviews, demographic data, rainfall data and consultations of health centre health registers, etc.). It emerged that the high prevalence of disease during flood periods is the result of flooding, which is itself caused by several factors (man-made and natural). Malaria is the disease most familiar to the population of the area, with a prevalence rate of 70% according to consultation records.

Key words: flooding, interaction, water-related diseases, N'Djaména, health risks, strategy.

GENERAL INTRODUCTION

GENERAL INTRODUCTION

Life of any kind is impossible without water. Whether for humans, animals or plants, water is essential. In the context of this study, the focus is on flood waters and their repercussions on populations, in particular the cases of water-related diseases that result.

Indispensable for life, water, particularly freshwater, has become a coveted resource in a world marked by over-consumption, and the focus of numerous territorial conflicts. While in economically developed countries access to drinking water of sufficient quality and quantity is a formality, this essential human need is not being met a global scale.

In fact, for years there has been increasing debate about flooding, which has now become part of our daily lives. Hardly a year goes by without Third World countries in general, and African countries in particular, crying flood. But so far, no concrete solution has been found, and the floods continue to ravage the world's inhabitants. They need to stand up as one to put an end to this problem, which may seem simple on the face of it, but the damage it causes is enormous.

Long considered a myth, the figures do not lie and the statistics are striking, as floods continue to ravage the world and more specifically the towns of Chad.

According to the records of the Office for the Coordination of Humanitarian Affairs (OCHA), flooding in N'Djaména over the past decade (20122022) should be nothing new to the local population. Heavy rainfall has been recorded since 1988, and ten years later in 1998. Although the 2013 rains claimed 20 lives and affected almost one and a half million people, they are still an old phenomenon. According to the same source, 94,211 houses were destroyed and hundreds of head of cattle washed away. According to statistics from the Ministry of Agriculture and Irrigation, 225,700 hectares of arable land were flooded.[1]

According to a press release from the Miandoum district, at around 2am on 12 June 2022, heavy rain accompanied by a violent wind and hailstorms hit five villages in the Miandoum district and sub-prefecture, specifically Ngalaba, Bendoh, Mamkéri, Meurmoeuel and Béka, causing huge damage to both people and property. Explicitly, the report shows 1027 households destroyed, 5011 victims, 2302 houses, 352 injured, 10 broken bones, 5 deaths, damaged food stocks and uprooted fruit trees. Many families have become homeless, and are crying out for help from the authorities, national/international organisations and anyone of good character who can come to the aid of these people in distress - that's the word that best describes the situation.

Floods have always been analysed in geography. This type of study has focused as much on physical aspects (river morphology, hydrology, climatic conditions) as on the impact of this phenomenon on human societies. Over the last two decades, there has been a resurgence of interest in flood research in geography, largely as a result of scientific and government concerns about climate change. Although geography is not the only discipline to take an interest in the impacts of climate change on river environments and societies in general, it does play an active role, particularly through the production of innovative work based on a variety of methodological approaches that straddle the disciplines of the natural sciences and the humanities. The phenomenon of flooding was addressed very early on by researchers who often combined both physical and social approaches. Human societies have always been concerned about the impact of floods on their environment. These events, often referred to as

[1] Courrier Afrique n°43 December/January 2013

natural disasters, have indeed caused a great deal of damage. While a number of geographical studies explore different aspects of the flood phenomenon, more specialised approaches are being developed, drawing on the hydrological sciences.

Flooding and water-related diseases are major challenges for many urban areas in Africa, particularly in Chad. The 7th arrondissement of the city of N'Djaména, the capital of Chad, faces these problems on a recurring basis, jeopardising the health and well-being of its inhabitants. This study aims to explore the interactions between flooding and water-related diseases in this specific locality, examining how these phenomena influence each other and analysing the responses and strategies adopted by the local population to cope with these challenges.

At the heart of this study are the health implications and socio-economic consequences of flooding and water-borne diseases in the 7th arrondissement of N'Djaména. Flooding can lead to contamination of drinking water, the emergence diseases such as malaria and diarrhoea, and a deterioration in the living conditions of local residents. Understanding these complex dynamics is essential if we are to develop effective strategies for preventing and managing these health risks.

With this in mind, this study aims to provide an in-depth analysis of the risk factors, health consequences and local practices for adapting to flooding and water-borne diseases. By giving a voice to the residents of the 7th arrondissement of N'Djaména, we hope to provide valuable information on the interconnected issues of water, public health and the environment in this specific region, thereby contributing to the development sustainable and holistic solutions to improve the quality life of the city's residents.

Specialists say that these recurring floods are a combination of several factors, both natural and man-made. Studies have been carried out water, sanitation and hygiene, but the results are bitter. More than 13,000 children under the age of 14 die every year from water-borne diarrhoea, mainly in Eastern Europe and Central Asia. In 2006, more than 170,000 cases of water-related disease were reported, and this is probably an underestimate. These health statistics reflect only part of the reality. (WHO, 2006).

The prevalence of water-related diseases has increased at an unprecedented rate in recent years. The WHO states that 80% of infectious diseases are caused by poor water quality and are one of the leading causes of death worldwide, accounting for 25,000 deaths a day (1/3 of all deaths)[2] , cited by Matembé (2015). In addition, flooding can indirectly encourage the contraction of vector-borne diseases such as malaria. Water-related diseases are caused both by dirty stagnant water and by "too much water", the latter aspect being of particular interest to us. In its report of 26 June 2008, the WHO estimates that dirty water is the cause of 9.1% of illnesses and 6% of deaths worldwide each year. Children are the first victims, since water is the cause of 22% of illnesses in children under 14 (Pierre et *al.*, 2012).

The inhabitants of the city of N'Djamena, the commune of the 7th arrondissement and certain neighbourhoods, in particular Abena, Atrone, Ambata, Amtoukoui and Boutalbagara, are not spared from flooding and the consequences that go with it, particularly in terms of health (water-related diseases). With this in mind, our research aims to analyse the strategies of city dwellers in the face of flooding and water-related diseases. We will also focus on the main causes of flooding, how flooding can be the cause of water-related illnesses in urban areas, and study the prevalence of these illnesses in order to provide valuable assistance to the

[2] http://www. eau-seine -normandie.fr/

administrative authorities of the aforementioned commune in managing natural and health risks.

Environmental issues appear to be difficult, not to say complex, to address, which creates uncertainty in many of the scientific studies carried out.

Our work will be divided into three (3) parts, each with two (2) chapters:

The first part is devoted to the context, the framework of the study and the methodology adopted for this work. The second part is devoted to elucidating the various causes of flooding and water-related diseases, based on our documentary and exploratory studies and field surveys. The last part is devoted to the perception of floods and water-related diseases by the population of the area, the dominance of one disease over another and the strategies to be adopted to deal with the resurgence of floods and water-related diseases.

A number of studies carried out by researchers have used different methodological approaches to determine what triggers flooding in order to propose resilience strategies to alleviate the problem. This study is not intended to correct the strategies but to provide a specific layer.

I PART

CONTEXT AND METHODOLOGY

PART ONE. CONTEXT AND METHODOLOGY

INTRODUCTION TO PART ONE

All research or studies aim to solve a given problem. Ours is aimed at solving the problem of flooding and water-related diseases in order analyse the reactions of local residents to this problem. Research is based on a number of essential or key elements, so in order to carry out this study it is important to follow a few scientific guidelines. We thought it would be useful to divide this first part into two respective chapters, the first of which is devoted to setting out the problem from a general point of view, reviewing the literature, and describing the epistemology of the science to our work belongs. The second chapter describes the words or concepts, the study area, the field and laboratory tools used in this work, and finally presents the methodology followed.

CHAPITRE 1. CONTEXT OF THE STUDY

The general problem raised by the study that of analysing the reactions of city dwellers to flooding and water-related diseases. It is also a question of making the link between flooding and water-related diseases. A number of studies have carried out into flooding and water-related illnesses, here and elsewhere. Doing research means following a thread of ideas. Our first chapter is based on a number of parts: the problem and literature review, the research objectives and hypotheses.

1.1. THE PROBLEM OF THE STUDY

Flooding is one of the natural hazards to which thousands of people around the world are exposed. As well as being at the heart of the major problems facing the world's inhabitants, floods are the subject of a number of studies that attempt to analyse them through the prism of methodological tools. Basically, two approaches share the monopoly of flood . The first approach postulates that floods are the natural manifestation of the world's environmental and geographical evolution. As such, they reflect and express the natural order of things. The second approach maintains that floods are the result human actions on nature. These natural impacts include the irrational exploitation of natural resources (excessive logging, bush fires, etc.) and "uncontrolled" land use and urban planning policies (occupation of hilly sites, archaic development, etc.). In the latter case, flooding becomes a result social action, a phenomenon whose causes are to be found in the socio-anthropological actions of the affected populations themselves. We can also talk about another intermediate approach that lies at the intersection of these two approaches: in other words, an approach that considers that as much as floods are of a natural nature, they are also linked to the impact of human actions on nature. On closer examination, flooding has become a "total social phenomenon", to use a socio-anthropological expression dear to Mauss (2013). In other words, flooding is the focus of a number of considerations combining health, social, housing and planning policies. They call for a number of different explanations: flooding as a natural phenomenon, flooding as a consequence of human action, flooding as an expression of the wrath of the gods and dead ancestors against the living, etc.

According to Ocha (2022, p1), in 2022, torrential rains and excess annual flooding on the Chari and Logone rivers caused flooding that severely affected the population. Out of the country's 23 provinces, 19 were affected by the floods, affecting 1.3 million people. The city of N'Djamena was affected by the rise in the Chari river, displacing 254,483 people (42,483 households) between October and the end of December 2022.

Kovacs et *al,* (2017, p4), floods nevertheless represent more of an economic risk due to the material damage they cause than a lethal risk for the population affected. A comparison with other types of natural disaster shows, for example, that floods cause far fewer deaths than earthquakes.

The material damage caused has a significant impact on the living conditions of people affected by flooding, particularly through the deterioration of buildings (homes and strategic infrastructures such as hospitals and schools), the disruption or breakdown of networks (drinking water, sewerage, waste, energy, transport), disruption or stoppage of economic activities, and the displacement of populations.

Several studies have shown that floods claim around 20,000 victims a year worldwide. The demographic explosion is exacerbating vulnerability both in human and material terms in certain areas.

Floods are among the natural disasters that most affect human societies.
In Chad, the general flood situation is critical, with enormous damage, including loss of human and animal life (livestock) and destruction of homes and crops.
Climatic disturbances with sudden variations rainfall intensity have had a significant impact throughout Chad, causing flooding in most localities.
According to Mbayam (2013, p15), in the post-independence period, after the 1980s, the city of N'Djaména experienced very rapid urbanisation, which had a significant impact on the water cycle. Soil impermeability has also increased tenfold, leading to a marked increase in run-off volumes. Flooding in the city of N'Djamena is becoming more and more frequent, and despite the new developments that have been undertaken, episodes of rainfall have just as negative an impact on the population.
N'Djamena's 7th arrondissement is the one that suffers from flooding every year. It's not a new situation, as residents complain every rainy season. The residents of some of the districts in the 7th arrondissement have been living in this situation more than a decade. For some journeys, others have to use dugout canoes, as the roads linking these neighbourhoods to each other and to the others are flooded every year by puddles. These roads can be around one (1) to two (2) kilometres long if you decide to go around them, so some people have to use the pirogues provided.
Water is the source of life, because life on earth is possible thanks to the existence of this vital resource, a commodity of great importance to living beings. According to the WHO (2014), approximately 2.5% of the earth's water is freshwater, which is largely inaccessible because it is the most important natural resource that makes life possible, supports ecosystems and human endeavours. Water is therefore both a strategic resource and the fundamental building block of a healthy economy.
There is a strong link between human health and water. It is often said that "water is life", but this same water is also sometimes the source of a number of problems.
One of the factors affecting health is polluted, poor-quality water, which is one of the causes of illnesses that affect human health.
As well as having material and economic repercussions, flood-related issues also have an impact on the health sector. Around the world, there is much more interest in hydrological risk management, which is also the case in Chad, but more often than not, the prevention aspect and strategies for dealing with flood-related diseases are overlooked.
In Chad, the year 2020 will be marked by a very heavy rainy season. Heavy rains in many provinces in recent months have caused widespread flooding across the country. According to the Chad Red Cross (CRT), 171,160 people had been affected by the floods by the end of October, and the number of people affected nationwide continues to rise .[3]
The flooding in Chad, particularly in some districts of the 7th arrondissement of N'Djamena, has had an impact on the health of the population. We can highlight the contamination of water quality, generating epidemic risks if the appropriate consumption instructions are not followed, and the soiling of rainwater stagnating in retention basins or unhealthy conduits (gutters) are sources of contractions of water-related diseases either through direct or indirect contact.
Water-related illnesses in children account for more than 35% of the top ten causes of consultation in N'Djamena. Malaria is the most common illness, followed by typhoid fever,

---

[3] Situation report OCHA (Office for the Coordination of Humanitarian Affairs), 2019

dermatitis, bilharzia and gastro-enteritis. Some 19,000 children die every year in Chad from water-borne diseases, particularly diarrhoea while one child in 5 suffers from it on a daily basis, said Célestin Traore, Head of the Child Survival and Development Programme at the United Nations Children's Fund (UNICEF) in Chad, in an interview with Anadolu. Urgent action is therefore needed on household sanitation and water management.

Water-related diseases continue to wreak havoc in the city of N'Djamena. These diseases are now considered to be a public health problem. Polluted water can be the source of serious illnesses such as dracunculiasis, bilharzia, malaria, cholera, typhoid fever, dermatitis, etc.

There are two reasons for choosing this subject.

Firstly, the perception of flooding and water-related illnesses in a number of neighbourhoods in the 7th arrondissement. The focus is on the main causes of flooding and the health repercussions.

Secondly, we need to find out about the most common types of water-related disease and their prevalence, then analyse the strategies adopted by local residents to deal with them, and finally suggest other strategies for strengthening their response.

The aim of this study is to analyse what people living in urban areas do when faced with the natural (flooding) and health (water-related illnesses) risks of flooding. We need to look at the factors that cause flooding, identify the different types of recurring illnesses that result from it, and understand how people deal with these phenomena. The study is based on the use of statistical (demographic, health) and rainfall data, which has enabled us to understand the phenomenon of flooding and the temporal and spatial dynamics of water-related diseases. To do this, the study will attempt to list the most dominant diseases in terms of percentage of the area. Urban management and the demographic explosion are also key factors in assessing the rate of change in water-related diseases before, during and after flooding.

1.2. LITERATURE REVIEW

Floods and water-related diseases have been the subject of several studies; however, our study consists of going further by highlighting what has not been said by the various authors we have consulted.

Various methods have been used as part of a diachronic study of floods and water-related diseases. In our study framework, it is important to study the interaction between floods, natural and anthropogenic factors, urban management and human health. This has enabled us to show how poor urban management (demographic pressure , land use and urban development) and natural factors (climatic variations and geomorphology) exacerbate flooding, which is the source of several water-related diseases in some districts of the 7th arrondissement of N'Djamena.

Caught between two schools of thought - Ritter's deterministic view that man does not depend on the environment and Ratzel's naturalistic view that man is a product of the environment or of nature - this work focuses on man's interaction with the natural environment.

Antoine et *al,* (2016, p187), since the 1980s, new themes have been emerging in a discipline that is in the midst of renewal: spatial justice, geopolitical issues of world division, sustainable development, global warming, new mobilities and migrations, gender discrimination... illustrate this profound evolution. Pioneering studies on risks, whether natural or social, are a clear reminder that all these phenomena are part of an ideological and historical context. A physical risk, such as a flood or volcanic eruption, is not just natural. Talking about it and studying it illustrates the relationship between man and environment,

human geography and physical geography.
Geography is a discipline that is often seen as a compilation of knowledge borrowed from the earth sciences, life sciences and other human and social sciences. Sylvain et al (2003):
"It is the only discipline that can claim the triple status of human science, social science and science of nature".
Human societies have always had to cope with natural phenomena. The risks they face are both natural, i.e. without human intervention, and man-made, i.e. linked to human activities. The degree of vulnerability to a natural hazard depends on population density. In geography, a risk is the probability that a hazard will cause damage to a human society. There are many different types of riskincluding hurricanes and cyclones. But what interests us here is flooding.
Indeed, managing the risk flooding in African cities, particularly in the 7th arrondissement of the city of N'Djamena, should be everyone's business. We are facing unprecedented and galloping urbanisation that is beyond the control of residents and local government authorities. Flooding problems are the intersection of two actions, namely climatic parameters and urban management.
Chad, like every other country in the world, has to contend with flooding every year. Several natural factors are involved in this phenomenon. These include reduced rainfall and unprecedented rainfall variations. With the late arrival of rains in recent years, but sometimes in large quantities, climate change is at the heart of debates around the world, particularly in the Sahel countries (G5 Sahel). According to Mbayam (2013, p15), geomorphology plays a very important role in the risk flooding, as the city of N'Djamena is known for its impermeability, which makes it difficult for water to infiltrate and causes it to circulate on the surface. The result is a rise in surface water levels, leading to flooding. However, in highlighting the issue of soil impermeability, we must not ignore the fact that water stagnating around points such as boreholes and wells cannot be infected. This is because, when the water overflows, the marshes fill up wastewater and flow into the water points, causing them to become contaminated.
In the second case, geography appears here as a construct; it is through new developments that man has been able to tame nature, and therefore space. By occupying the very space that was intended for water retention, man's actions often lead him to live in the bed of watercourses, which causes flooding, because the possibility of building drains is now limited. In addition, climatic parameters play a very important role in flooding. Given the many complex problems linked to sanitation and flood management facing the city, flooding in the city of N'Djamena is becoming increasingly frequent, despite the new developments that have been undertaken. Rainy periods also have a negative impact on the population. The impermeability of the soil has led to a marked increase in the volume of run-off. Controlling urban development and managing rainwater is becoming a major challenge for local players, who are responsible for urban development and are keen ensure the comfort and safety of their fellow citizens. Little is known about the circulation of water disrupted by man. Buildings increasingly occupy areas reserved for water: riverbanks and riverbeds.
Tchotsoua & Bonvallot (1993) in their article on the phenomena of erosion and urban management in Yaoundé stated that the unprecedented growth of the population in urban areas, occupation in non-compliance with the town planning scheme and various development exacerbate the risks in these areas even further.
Density, population growth, urbanisation and its management, as well as poverty, exacerbate

vulnerability to flooding. The 7th arrondissement is hemmed in by marigots. These areas are gradually being occupied by man, leading to spatial growth. This water, forced by man, leaves its bed and flows towards homes and other buildings. In 2013 and 2014, the construction company ENCOBAT began work the construction of drains and retention basins; this development work has not really helped in the fight against flooding. Given the width and depth of these hastily constructed drains and retention basins, as soon as the first rains fall, the inhabitants are already engulfed by the waters.

In recent years, N'Djamena has received a large amount of rainfall, in contrast to previous years. This sudden change is one of the causes of flooding in the city of N'Djaména, particularly in some districts of the 7th arrondissement: variations in climate and rainfall.

In order to understand the causes of flooding in the Abena, Atrone, Ambata, Amtoukoui and Boutalbagara neighbourhoods and to participate in its management, it is important to know the hydrological behaviour of the watercourses and the characteristics of the terrain. On the basis of these processes, we can better propose measures to combat flooding and the diseases that go with it, in order improve the living conditions of the inhabitants of these neighbourhoods. Bakari (2018):

It's time for the inhabitants of our African countries to practise a culture of risk, which prepares the population psychologically for the effects of hydrological and health risks, so the problem can be tackled at an early stage.

For Antoine et *al* (2016, p195), applied geography deals with disasters as the result of a failure to conceptualise risk or a failure to plan and anticipate. It approaches human developments as constructs designed to minimise risk but, in so doing, may produce new risky practices. It draws up typologies of major risks, potential vulnerabilities and their severity, in order to develop programmes of preventive measures tailored to the frequency and magnitude of the phenomena. In terms of both natural hazards and man-made impacts, the scientific approach is characterised by three stages : identification and nomination of a risk situation, the probability of its occurrence, and the consequences for society in space and time. This scientific assessment, which is often delicate in conditions of uncertainty, is intended to enable better risk management.

Studies focusing on health have gradually taken their place within the social sciencesleading to the emergence of geography through a number of research projects. In France, Max Sorre, the geography of health is now at the crossroads between the geography of disease and the geography of care. For a long time, health geography has studied illnesses and their causes, using cartography as a tool to spatialise them.

Bétinbaye et al (2015, p2) argue that in urban areas in general, and in Chad's largest cities in particular, rivers are, along with groundwater, the main sources of water storage and supply for the population. The River Chari and its main tributary, the Logone, supply water to three of Chad's four main urban areas, namely N'Djaména, the state capital, Moundou, the main economic town, and Sarh, the first town to be created in the country.

A number of authors have come to the conclusion that most water-related illnesses are caused by problems with urban sanitation and the quality of the water consumed by the population.

Nguedjo (2011) conducted a study on geomatics and the analysis of sanitation and health risks in an urban environment: the case of the Mbanya catchment area in the city of Douala. The overall aim was to use geomatics to assess the state of sanitation and its consequences in Mbanya with a view to proposing measures for its improvement.

For Jean (2007, p5), water-related diseases, in particular parasitosis, are limits imposed by nature on the expansion of human societies around aquatic environments. These parasitoses form pathogenic complexes whose different elements are man, the causal agent and its vector or intermediate host, and each element is subject to the constraints of the aquatic environment. According to Razak (2017), the proliferation of waterborne and water-related diseases in Ngaoundéré is a cause for concern, with 40% of diseases being transmitted, either by germs contained in the water, or by vectors that stay in the water or use it for their reproduction. His study showed the disparity in the distribution of water-borne diseases in Ngaoundéré. His study showed that certain health areas in our study zone have high prevalence rates of water-borne and water-related diseases, and that these high prevalence rates are justified by a number of explanatory factors. The identification of these factors has made it possible explain the vulnerability of the health areas in the study zone.

CHAPITRE 2. EPISTEMOLOGY, CONCEPTUAL FRAMEWORK, THEORETICAL, GEOGRAPHICAL AND METHODOLOGICAL FRAMEWORK THE STUDY

A study in geography is based on several theories: some specific to geography and others borrowed from the social sciences. It is also important to make a scientific study of the discipline in which our study falls, to know the meaning of the different words or concepts used in the work so as to adapt them better to our field of study. The meaning of each word or concept varies according to the field in which we are working and the disciplinary field in which we are involved. The study area is representative of a geography subject because the geographer studies the phenomenon in time and space. Essentially, this chapter focuses on the various parts listed above.

2.1. EPISTEMOLOGY, CONCEPTUAL AND THEORETICAL FRAMEWORK

2.1.1. Epistemology of health geography

This research has not only made it possible to determine the influence of urbanisation on the increase in run-off water, but also to highlight the impact of urban activities on the quality of this water and the resulting impact on receiving environments. This has led to new concepts in urban rainwater treatment.

Cities have never been able to grow and develop far from water, which is as necessary for biological life as it is for the economic life of large numbers of people. What is new in modern times, both in developed and Third World countries, is the increasing speed of urbanisation and its ever-growing need for space.

Our study borrowed the methodological approach of health geography to explain the interactions between flooding and water-related diseases in Chad.

Sébastien et *al* (2022) argue that health is a human, social, political and economic issue for every country and society. It was therefore logical that geographers who were more sensitive to people than to spatial forms should make it one of their research themes. Although health geography is a traditional sub-discipline of French geography, it is becoming less and less homogeneous. When applied to the field of health, the indicators change, but the geographer's approach and objectives remain the same: the aim is to characterise a society through the geographical study of its health.

As the offspring of urban geography, the geography of health sheds light on the links between social classes, living spaces and the health of the urban population. In the nineteenth century, a doctor by the name of J. Snow (1855) measured the location of patients suffering from the cholera epidemic in London in 1854 and was able to find the polluted water well that was the source of the contamination. His method can be described as scientific. Public health research soon went beyond location measurements to include social and spatial inequalities, at a time when hygienism was becoming an element of urban planning. Geographers working in fields close to epidemiology were opening up to studies on the effects of poverty, lack of education and marginalisation on public health. Under the influence of behavioural and radical visions, health geography was born.

Antoine et *al,* (2022, p181-182), one of the most accomplished studies is that by P. Gould (1993), *The slow plague,* on the spatial spread of AIDS, on an international, national (United States) and urban scale: spread from central contact points to the residential outskirts; spread along regional transport routes; global spread as a result of air travel. Since the 2000s, this geography of health has become more diversified: the influence of poverty on pathologies, migration, education, the environment, etc. Often mapped out in health atlases using GIS, this shows not only the spread of diseases, but also the unequal distribution of care (policy, health services, hospitals, etc.) and their role in the well-being of populations.

As regards the impact of environmental variables on health, for example, when we look at the epistemology of health geography, we quickly come up against a complexity linked on the one hand to the parallel history of the health sciences and geography since Hippocrates, and on the other hand to the multiplicity of health determinants. The term "medical geography" is attributed to a German naturalist, Finke, and dates back to 1772.

Picheral (2001), classical medical geography encompasses two more or less independent fields: the geography of diseases or medical ecology, and the more recent geography of healthcare. The foundations of modern medical geography are to be found in the work of Max Sorre, who developed the notions of pathogenic spaces - or rather complexes - by integrating disease into its environment, or rather by reintegrating it, since Hippocrates had already insisted on the need to contextualise the medical act. In those days, doctors were nomads. It was therefore very useful for them to have an idea of the pathologies they would encounter in a particular place.

For Sébastien et al (2022), the approach proposed by Max Sorre has paved the way for numerous studies which essentially consist of using statistics to establish correlations between the incidence of certain diseases, mortality rates and the fact of living in certain places. This approach to medical geography therefore converges in many ways with epidemiology, and this probably has more to do with the history of geographical thought than with Hippocratic principles.

The geography of risk, by focusing on the images of risk, its social valorisation, its dynamics and its consequences, studies it in a global way, both in its representation and in its territorial management. From a physical geography, it becomes a behavioural and critical geography.

On the other hand, it was not until the last decades of the twentieth century that individual and collective behaviours were considered as a consequence of the social environment, both in the list of health determinants and in geographical analysis, and above all that they were considered alongside the physical determinants in an overall environmental concept.

2.1.2. Conceptual and theoretical framework

In order to have a common understanding of the terms used in this work and to prevent any risk misinterpretation when reading, it is important to define certain major concepts. More than any other, the analysis of a geography topic must necessarily be based on a prior conceptualisation effort that must ensure that it is put into context. Our analysis therefore based on the following fundamental concepts:

Interactions

In a more general context, an interaction refers to a dynamic or interdependent relationship between two elements, variables or entities, where the effect of one is influenced by the presence or level of the other. Interactions can manifest themselves in different ways depending on the field of study or the context considered, and they are often complex and multidimensional. In the social sciences, political science, psychology, physics, biology or even computer science, interactions can be observed between various factors, such as individuals, natural phenomena, components of a system, statistical variables, and so on. Interactions can be direct or indirect, simultaneous or successive, linear or non-linear, positive or negative. Identifying and analysing interactions is essential for understanding the interdependent relationships between the different elements of a system and for predicting how these relationships may influence the results or phenomena observed. Interactions can play a crucial role in understanding the complex processes, feedback loops, synergistic effects or interdependent situations that characterise many natural and social systems and phenomena. In the context of the link between flooding and water-related diseases, an interaction could manifest itself as follows: When floods occur, floodwater can contaminate drinking water sources by carrying pathogens (bacteria, viruses, parasites) and various contaminants. This can lead to a significant increase in water-borne diseases such as diarrhoea, cholera, leptospirosis and typhoid fever among the exposed population. In health geography, interaction refers to the complex relationships and processes that influence the health of populations as a function of their geographical environment. This discipline studies how geographical factors, such as location, spatial distribution, landscape characteristics, accessibility to health services, land use, etc., interact with social, economic, cultural and environmental factors to shape the health of individuals and communities.

Flooding

A flood is a temporary submersion by water, which may be rainwater, used to characterise the overflowing of a watercourse or a water retention basin. In the National Survey Glossary (OIE, 1995), floods are hazards associated with flood-prone areas: harm to life, health or damage that may occur in such areas. Floods can be water-related hazards or inconveniences to which people are exposed, or major water-related disasters faced by populations. There are several levels of risk, depending on the severity of the damage to be feared in terms of submersion, the speed of the current and the vulnerability of the sites exposed. Floods occur when a body of water covers land that is normally dry. They are often caused by extreme weather events such as heavy rainfall, storms, snowmelt or flash floods. Floods can also result from the failure of dams, dykes or other hydraulic infrastructures.

Water-related diseases

According to WHO (2014), disease is any alteration in health status usually manifested by abnormal sensations. Thus, we can define water-related diseases as any illnesses caused by ingestion or contact with unsafe water. In its etymological sense, the term disease comes from the Latin *mal habitus*, meaning in poor condition. In a general sense, the term water-related diseases refers to illnesses (and by extension health risks) linked to the quality of water (from flooding). Water-related diseases are illnesses caused by drinking water contaminated with pathogenic micro-organisms, toxic chemicals or other harmful agents. Contaminated water can come from a variety of sources, such as polluted surface water, untreated waste water, contaminated wells, faulty water distribution systems, etc.

Theoretical framework

Every science needs a theory, because it is through the theory that the tools are forged to support the thread of thought. We have drawn on a number of theories from the human sciences, particularly geography.

Systems theory (cybernetics) or catastrophe theory can be used as a guide to understanding cross-disciplinary phenomena. It seeks to account for the functioning and malfunctioning of systems that generate crises and risks. Haimes (1998) proposes methods for analysing technological risks based on systems theory. We will use them to study the factors exacerbating flooding in the Abena, Atrone, Ambata, Amtoukoui and Boutalbagara districts.

Cindynics is largely based on the study of complex systems, the complexity of which is accentuated by the multiplicity of interactions that unite the different components a disaster. According to disaster theory, a disaster occurs when a continuous variation in causes leads to a discontinuous variation in effects. Disasters are therefore prepared on a continuous basis over a long period of time suddenly erupting. It should be noted that the discontinuities analysed and modelled by mathematicians are not necessarily disasters that cause casualties and damage in the usual sense.

Determinism: no discipline has shown as much enthusiasm as geography for the study of spatial facts and the apparent relationships between these facts. This has led traditional regional geography to pay special attention to one particular aspect of these relationships, namely the search for direct causalities between physical and human phenomena. Apparent causalities are thus highlighted, without taking into account the subjectivity of the researcher and the partial nature of his observations. According to Antoine et al (2016, p47-48), in all cases, one constant dominates: the direct influence of a factor (the environment) on human actions. For example, the presence of water resources influences the location of villages, soil types influence agrarian structures, and sites influence urban settlements.

2.2. GEOGRAPHICAL FRAMEWORK AND METHODOLOGY

2.2.1. Geographical context and reasons for choosing the study area

The 7th arrondissement is one of the 10 arrondissements of N'Djamena, the capital of the Republic of Chad. With a surface area of 108.752 km^2, the 7th arrondissement is made up of 15 districts according to the new administrative division (Figure 1). Our study area includes a number of districts, namely: Abena, Atrone, Ambata, Amtoukoui and Boutalbagara. N'Djamena is located in the hot, dry Sahelian zone. It lies between 12.11° north latitude and 15.04° east longitude. Situated at the confluence of the Chari and Logone rivers, it is bordered to the north by the Mani sub-prefecture, to the east by the Linia sub-prefecture and to the south-east by the Logone Chari sub-prefecture. It has a relatively flat, clayey topography. The central part of this relief is in the form of a basin with very little difference in level.

In this region, the prevailing climatic conditions are quite restrictive for the development of fauna and flora. Since independence, the city of N'Djamena has undergone rapid urbanisation, which has had a significant impact on the water cycle.

In general, little is known about the problems associated with water in this environment. It is with this in mind that this work has been undertaken, based on the observation that many people are victims of water-related illnesses.

Why choose just a few districts?

The 7th arrondissement is one of the largest in N'Djamena. Selecting a few districts allows us to study the phenomenon in areas under the influence of flooding. Similarly, areas that are not overly exposed to flooding are studied in order to assess the incidence of water-related diseases. The Abena, Atrone, Ambata, Amtoukoui and Boutalbagara districts are the most exposed to flooding. Neighbourhoods such as Dembé and Chari mongo, although less affected by flooding, are not without their share of water-related illnesses. Geomorphology is one of the factors exacerbating the hydrological risks in these neighbourhoods, as N'Djamena is characterised by its impermeability (with little infiltration). Added to this is climatic variation. The Abena, Atrone, Ambata, Amtoukoui and Boutalbagara neighbourhoods are landlocked, and new developments are being carried out. The uncontrolled occupation of certain areas that were once intended to collect water now results in water being discharged into households, contaminating wells and boreholes and affecting the health of the population. In neighbourhoods not affected by flooding, the mobility or migration of residents for a variety of reasons (markets, family visits, etc.) must also be taken into account. Lastly, the issue of unhealthy living conditions should not be forgotten, as we have focused on hygiene and sanitation.

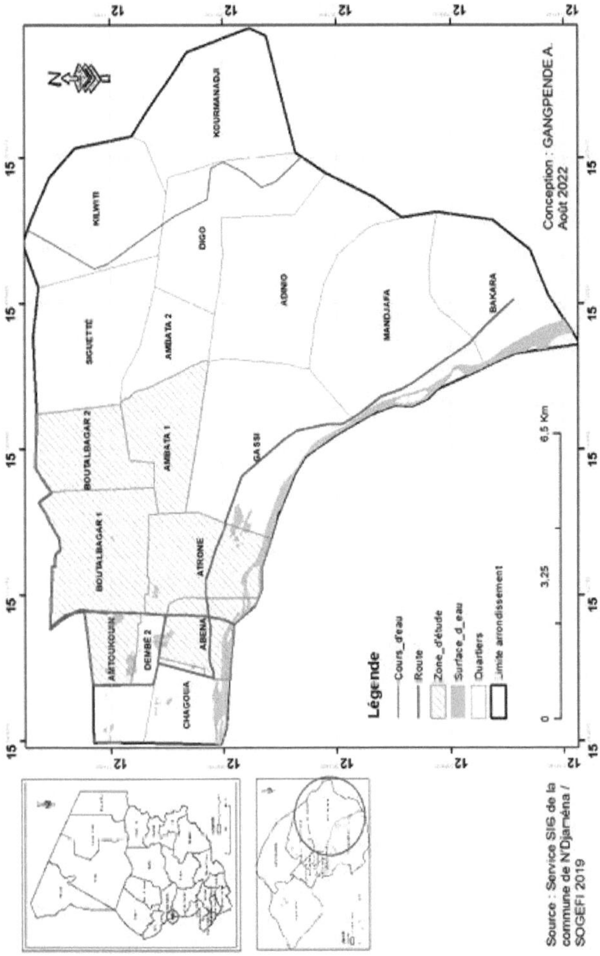

Figure 1. Map of the 7e arrondissement municipality

2.2.2. Tools and methodology

Practical framework

Our practical framework summarises a large number of tools and techniques that were used to collect empirical data. In particular, it includes documentary research, which enabled us to focus on our problem, direct observation and interviews and questionnaires as investigative , which used to gather information from resource persons. Our study is based on a quantitative and qualitative approach.

Direct observations in the field

As Tamira (2023, p25) points out, unlike participant or participatory observation, direct observation reduces to a minimum the researcher's ability to *intervene* in the actions and interrelations of the individuals under study. Observation thus becomes a technique for capturing unprecedented human realities, in action but always elusive. Direct observation provides research with solid data for analysis, giving it an empirical character and the only solid basis for human knowledge. However, the researcher's observation, beyond its *intrusion* into the life of a given social group under study, must always be objective and divorced from the social, cultural and even intellectual origins of that group.

For the purposes of this work, we opted for sequential observation of the facts, i.e. one-off field visits, each of which involved observations, note-taking and, in short, a partial investigation. This approach is both essential and practical, in that it has enabled us grasp and pinpoint the facts directly from our direct observations of the discrepancies.

Qualitative component: data collection and analysis tools

The qualitative tools used include individual interviews and Focus Group Discussions (FGD).

Interviews

An interview is a conversation with a person in which he or she is asked about his or her life, ideas and projects, with a view to either publishing or disseminating the content, or using it for analysis. We therefore used a structured interview. A number of interviews were conducted as part of this work, and it should be noted at the outset that these were semi-directive and free interviews.

Open interview

The aim of the open-ended interview is to give the interviewees the opportunity to tell us a great deal about their lives and their relationships with the external players present in the area thanks to the oil company. It is intended, and this is the whole point, to be a means of reconstructing the history of the area, based on life stories and individual biographies. Unexpected questions may be asked of the interviewer.

Semi-structured interview

The semi-structured interview is a qualitative data collection technique which, like the unstructured interview, puts the researcher in contact with the interviewee. Here, unlike in a free interview, the respondent has only a limited amount of freedom, which is more or less controlled by the interviewer. But completeness is not ruled out if the respondent stays within the already predefined framework of the question. The semi-structured interview is of twofold interest in this work. On the one hand, it allows us to compile data in a very specific area of research, based on interviews with respondents.

Analysis toolscontent analysis

The qualitative data collected in the field was analysed for content. The triangulation method is also used here, as it offers a better possibility of analysing the information gathered by comparing its different sources. If systematic recording of social facts is impossible, the triangulation method is a better guarantee of the reliability of the data collected. Analysis method adapted from (Tamira, 2023).

Content analysis

Analysis is a measurement tool, a technique for objectively describing and interpreting empirical data. In this study, our monographic field data were analysed and interpreted in the form of statistical tables, diagrams and histograms.

Quantitative aspect

The quantitative aspect of our work uses the tools of the quantitative method in social science research, particularly in global geography.

The quantitative data collection tool

The main instrument of quantitative research is the questionnaire. It is a statistical method for quantifying the information gathered. It makes it possible to reach a large audience in a relatively short space of time, unlike qualitative research, which is more demanding and requires a longer physical presence with the respondents.

The questionnaire

The questionnaire, more than just a collection of questions, is a tool for collecting quantitative and quantifiable data. It makes it possible to survey a vast population and collect a large amount of data. There are a number of elements involved in this questionnaire stage. There is the choice of survey site and the number individuals to be surveyed. The choice of survey site includes the neighbourhoods experiencing the phenomenon. This series leads on to the household surveys, which gave us an overview of the situation in the field and the perceptions of the local population, health workers and local authorities. This section enabled us make contact with the population and the health authorities, from which we drew on the consultation and care registers to see the rate change in our study area.

Table 1. Structure of the questionnaire

Number of the Section	*Section title*	*Number of questions per Section*
Section 1	**Identification of the file**	*4*
Section 2	**Identification of the informant**	*9*
Section 3	**Perception of flooding problems and water-related illnesses**	*26*

Source: Target population survey questionnaire

Statistical analysis tools: SPSS and EXCEL

The quantitative data collected as part of this study were analysed using two software packages, Statistical Packages for the Social Sciences (SPSS) and EXCEL. The results of this analysis are presented in the form of statistical tables, percentages or figures (diagrams). These quantified data are added to the qualitative data to give a complete analysis of this work.

The sample or study population

Having indicated the survey methods used and their data collection tools, it is now time to determine the sample concerned by this work.

Sampling

The target population is the set of subjects of interest to the researcher, to whom the latter would like to be able to apply the results of the study or extrapolate the conclusions. The study population is the inhabitants of the 7th arrondissement of N'Djaména, in particular the target neighbourhoods. To draw the sample for this study, we used the non-probability sampling technique. This type of sample makes it possible to select elements using non-random methods. There is no way estimating the probability of each item being included in a non-probability sample, and each item does not have an equal chance of being included. Sampling is a statistical method of calculation that is indispensable to studies in the human sciences. It is a process that makes it possible to determine a standard set representing the entire or global population. Sampling a given population means using learned techniques (probabilistic and non-probabilistic) to extract from it a representative fraction that will serve as the standard set for the study.

Mapping the areas affected

Using demographic data, we were able to draw up urban expansion maps for our study area, enabling us to see how space has been occupied over time. The coordinates collected enabled us to draw up a distribution map of the health centres in our study area.

Cartography is an indispensable tool in geography research, and even more so in the sciences studying risks of all kinds. It was extremely useful in the work we carried out.

II PART

CAUSES OF FLOODING AND WATER-RELATED DISEASES

PART TWO: CAUSES OF FLOODING AND WATER-RELATED WATER-RELATED DISEASES.

Flooding is a major risk, causing significant damage, and its management remains a real problem not only for Chad, but for sub-Saharan Africa in particular and the world as a whole. Rainwater has been the cause of a great deal of destruction, and the health aspect is always put to one side, if not forgotten, but it never ceases to wreak havoc around the world. To shed light on these problems and understand the mechanisms behind flooding and its health consequences in the 7th arrondissement of N'Djamena, in particular the Abena, Atrone, Ambata, Amtoukoui and Boutalbagara neighbourhoods, we have divided this second part into two (2) chapters explaining how nature and man can exacerbate flooding phenomena, while showing how flooding can be the cause of the development of water-related diseases in the Abena, Atrone, Ambata, Amtoukoui and Boutalbagara neighbourhoods.

CHAPITRE 3. RECURRENT FLOODING IN THE ABENA, ATRONE ABENA, ATRONE, AMBATA, AMTOUKOUI AND BOUTALBAGARA DISTRICTS BOUTALBAGARA: A COMBINATION OF NATURAL NATURAL AND MAN-MADE FACTORS.

N'Djamena, the political capital of Chad, is now a province thanks to the decentralisation policy. Its origins date back to 1973, from Fort-Lamy to N'Djamena, which means "we have rested", in other words, putting an end to colonial servitude. The city is divided into ten (10) arrondissements, comprising around sixty-four (64) neighbourhoods. The 7th arrondissement was created by Decree n°227/PR/MISD/99 on the restructuring of the municipal arrondissements of the commune of N'Djamena on 19 June 1999, and became the commune of the 7th arrondissement following the local elections of January 2012. Characterised by a Sahelian climate, the climate of N'Djamena, by two (02) masses namely the Harmattan and the Monsoon. The barrier between these two masses constitutes the intertropical front, which moves in rhythm with the dry season and the rainy season. The rainy season begins in June and ends in October, while the dry season begins in November and ends in May. It is characterised by the cool period from November to February and the hot period from March and ends with the start of the rainy season. The waters of the River Chari and those of the marigots in the 7th arrondissement cause flooding in the district during periods of heavy rainfall (Ndilbé, 2018). This chapter looks at the main causes of flooding. It then looks at the impact of natural and anthropogenic factors on flooding in the Abena, Atrone, Amtoukoui, Ambata and Boutalbagara neighbourhoods.

3.1. NATURAL CAUSES OF FLOODING

Natural causes all manifestations of nature without human intervention. Several aspects of natural origin come into play in the flooding process at Abena, Atrone, Amtoukoui, Ambata and Boutalbagara. These are: climatic variation and geomorphology.

3.1.1. Climate variation

Hydrography is the totality of the waters of a region, country or district. In its broadest and most common sense, the word "hydrography" refers to the description of the waters of the earth's surface, whether marine or continental. In the latter sense, hydrography aims to describe and explain the characteristics of organisms defined by [4] continental waters, whether stagnant or flowing.

There are several closed or endoreic basins in Africa. The two main ones are Lake Chad and the Okavango. Lake Chad, shared by Chad, Niger, Nigeria and Cameroon, is essentially a vast swamp where the Ouambé, from the west, and above all the Chari, from the south-south-east, are lost. The latter, which is by far the most abundant, forms, with its tributaries and various branches, a very complicated network.

Lake Chad, which varies in size from year to year, although it is constantly shrinking considerably, used to flow into ? the Bahr-el-Ghazal through the Bodélé lake depression. Today, this river and lake have dried up. Around the Great Lake, there are other smaller lakes that, like it, have no outlet. Lake Fittri, fed by the Salamat from the Ouadai, is certainly independent. It seems that in the past, rivers flowed down from the mountains of Tibesti and Air as far as Lake Chad. [45]

Hydrographically, the central part of Chad includes districts *such as* Abena, Atrone,

[4] https://www.universalis.fr/encyclopedie/hydrographie/

[5] http://www.cosmovisions.com/Afrique-Hydrographie.htm

Amtoukoui, Ambata and Boutalbagara, most of which are made up of marigots and a few streams. These marshes thrive on rainwater. When these marigots overflow their banks, the damage is enormous, causing recurrent flooding. As every source of water has its positive and negative sides, these waters, as well as enabling the inhabitants of these districts to engage in certain income-generating activities such as market gardening, fishing, etc., also have a negative impact on the health of the population.

N'Djamena has a Sahelian climate, with less than 900 mm of rainfall a year and an average temperature of 35°C all year round. Maximum temperatures can reach 50°C between March and May. The rainy seasons are short compared to the length of the dry season. The heaviest rains fall between July and August, and even September.

From November to February, the temperature fluctuates between 24°C in the evening and 34°C during the day, when it is cool. March, with its high temperatures of around 50°C, marks the hot season, which only ends when the rains begin. The rainy season usually starts in June (rarely in May) and ends in October, rarely in November. Every rainy season our study area continues to suffer material and human damage. These include the inaccessibility of roads, the destruction of homes, the resurgence of water-borne diseases, etc.

Floods are more frequent in our study area, their timing has changed and the relationship between them and climatic and rainfall variations has been reversed. In the past, regular rainy seasons did not cause enough risk of flooding, but in recent years we have seen an irregular rhythm of temperatures resulting in unprecedented rainfall. This climatic/pluviometric variation is one of the main causes of increased water levels in the city of N'Djaména.

Variations in rainfall mean more flooding in the capital N'Djamena. In addition to these irregular increases in rainfall, there are also frequent rises in water levels, a phenomenon that most often occurs at the end of the rainy season and is triggered by the overflowing of water. Every year, several districts of N'Djamena are affected.

The flooding in the Abena, Atrone, Amtoukoui, Ambata and Boutalbagara districts is linked climatic variations. It has been noted that the meteorological services do not really control the weather as they did in previous years, due to contradictions between the meteorological agency's forecasts and rainfall variations.

Table 2. Rainfall data over 20 years (2000 to 2019)

Year	January	February	March	April	May	June	July	Aoiit	September	Ottobre	November	December	Cumulative
20(111	0	0	**d**	0	59.8	90,8	249.7	196,8	40.4	39.1	0	0	676,6
2001	0	0	0	13	18.1	45,8	135.9	162,6	250,7	0	0	0	**6143**
2002	0	0	**0**	0	0	42,4	122.1	147,4	151.8	80,3	0	0	**544**
2003	0	0	0	0.4	109.3	51,1	113,6	291,8	81.1	16,7	0	0	**664**
2004	0	0	0	38,7	55,1	52,8	175,5	142,9	51,6	27,6	0	0	**544.2**
2005	0	0	0	0	15,3	53,6	170,7	190,4	79,9	10,3	0	0	**520.2**
2006	0	0	0	1	31,7	87,4	174,8	295,2	83.5	37,6	0	0	**711.2**
2007	0	0	**(1**	0	10,5	106,9	218.5	180	72.5	29,4	16,5	0	**6343**
2008	0	0	0	0	44.7	50,5	215.8	297,8	38.5	**8**	0	0	**6553**
2009	0	0	0	0	0	12,7	145	232,4	111,1	54,1	0	0	**555.3**
2010	0	0	**(1**	0	1.8	51,5	215.3	122,6	155,7	20,7	0	0	**567.6**
2011	(I	0	**(1**	0	10.8	77,2	59.6	180,6	156.2	0	(1	и	**484.4**
2012	0	0	0	0	30,1	53,8	154,3	266,9	94,3	13,8	0	0	**6133**
2013	0	0	0	0	3,3	26,6	140.5	249,9	86.1	1,7	o	II	**508.1**
2014	0	0	0	1.9	28,4	4,6	173,1	189,8	114,1	17,1	0	0	**529**
2015	0	0	0	0	0,9	81,2	243,4	200,6	81.9	1,1	0	0	**609.1**
2016	0	0	0	0	25,8	47,9	211.2	210,4	137,7	25,2	0	0	**658.2**
2017	0	0	0	0	27.8	49,1	153.9	134,4	118.1	16.2	0	0	**499.5**

2018	0	0	(1	30.5	61 58,7	300.7	227,3	64.4	11,9	0	0	**754.5**
2019	0	0	0	0	653 58,1	180.4	141,3	128.7	59,1	0	0	**632.8**

Source: National Meteorological Agency (ANAM)

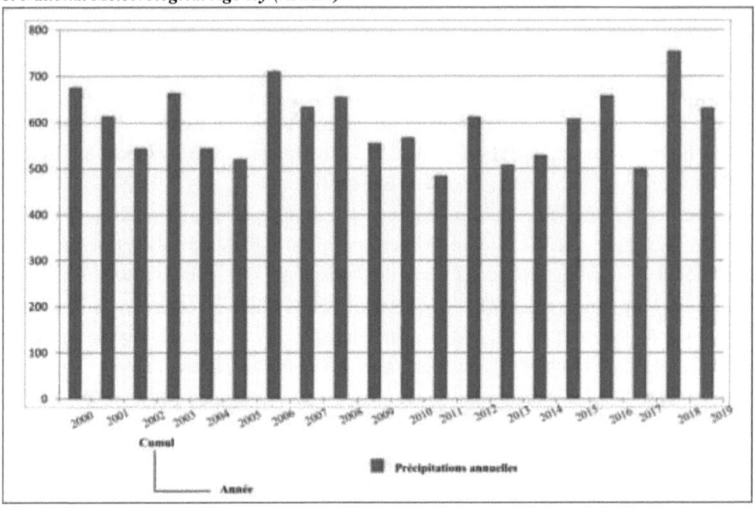

Figure 2. Variation in rainfall over 20 years (2000 to 2019) ***Source: National Meteorological Agency (ANAM)***

Table 2 and Figure 3 show the annual rainfall recorded in N'Djamena over a twenty (20) year period from 2000 to 2019. Rainfall has been irregular over this period, with 2006 and 2018 being the heaviest years on record. Climate variation has a perceptible influence on several variables associated with water, contributing to flooding.

It should be remembered that during heavy rainfall, marigots, water retention basins and many other drains overflow (in neighbourhoods with drains), and in the absence of adequate water drainage networks and poor infiltration (a phenomenon that occurs when the soil is impermeable), heavy rainfall is the trigger for flooding in our study area.

3.1.2. Geomorphology: the determining factor

According to the *Dictionary of Geography* (Pierre & Fernand, 2006), geomorphology is defined as "the *scientific study of the forms of the earth's surface and their evolution*". Often confused with geology, geomorphology is nonetheless a scientific discipline in its own right, whose object of research is the planet's landforms as a whole. From the Greek *gê* meaning earth, *morphê* meaning form and *logos* meaning discourse, geomorphology can be defined as the science that describes and explains the forms of the earth's relief (all the inequalities observable on the earth's surface). In this sense, the discipline lies at the interface of geography and the earth sciences. It plays an important role in regional and landscape planning procedures, as well as in the prevention of natural hazards and the exploration of natural resources. It is for this reason that it is adopted by town planning in the study of the physical form or structure and layout of a town.

The topography of our study area is a less rugged flood plain with clay soil. If it is located on a downstream slope, the upstream slope (river Chari). Clay or silt soils, which have a dense, poorly aerated structure, are impermeable soils. Impermeable soil lets little or no water through. As a result, rainwater is often stagnant on the surface (poor infiltration and surface run-off), which poses the problem of drainage.

The physical aspect, more specifically morphology, has a major influence on flooding. In recent decades, a number of geography studies have focused on flooding issues. Our study extends this scientific perspective. Some governments, aware of flooding phenomena, are focusing their policies on the future of the world in the face of climate change. In his work, Pardé (1959) was interested not only in the hydrological or climatic dimension of floods, but also in their consequences and risks for the populations affected, paying particular attention to damage they cause.

While many studies in geography focusing on hydrological problems are concerned with different aspects or factors of the problem, the main focus is on the shape of the landform and the speed and direction of flow.

The Abena, Atrone, Ambata, Amtoukoui and Boutalbagara neighbourhoods are located at a low altitude, which means that they very often suffer from flooding problems. Because they are located on a slope, these districts regularly receive water from other districts. The fact that they are so isolated is a crucial problem when it comes to getting the water to the river, as the water doesn't go anywhere because it is stagnant and quickly overwhelms the population.

Added to this is the issue of soil impermeability. The city of N'Djamena is notorious for its impermeability, resulting in poor infiltration of rainwater.

It should be noted that for flooding to occur infrequently, the quantity of water infiltrated must be greater than the quantity of water circulating on the earth's surface. Added to this are the intensity and abundance of the rainfall mentioned above, the lack of vegetation in the soil to limit the action of the water and the slope of the land (basin).

The neighbourhoods of Abena, Atrone, Ambata, Amtoukoui and Boutalbagara are surrounded by *Boutas*[6] . According to the story told by a former resident of Fort-Lamy (formerly N'Djamena), N'Djamena used to be called Bassin (Basin) because the surface area was practically covered by water, especially during the rainy season. It was only gradually that the inhabitants began to build, but as the saying goes, *"water always takes its bed even if it is forced to change its course"*. Nevertheless, the traces are still visible in the marshes that enclose the neighbourhoods.

Relief is an essential factor that largely determines the extent of run-off, infiltration and evaporation. It also characterises the hydrological nature of the land.

3.2. MAN-MADE FACTORS IN FLOODING

Flooding in the Abena, Atrone, Amtoukoui, Ambata and Boutalbagara neighbourhoods is a combination of two factors: natural and man-made. We have found that, alongside natural causes such as climatic variations and the shape of the land, human action is undoubtedly one of the most striking causes of flooding. These man-made causes include demographic pressure, land use and urban development.

3.2.1. Demographic pressure and land use

For several decades now, the city of N'Djamena has had to contend with flooding. The causes include rainfall and land use. Whether this anarchic occupation is partial or total, it has altered the water bed, leading to recurrent flooding. Demographic pressure, which is poorly managed by the authorities in charge of the municipality whose districts we studied are devoted to it, partly explains the flooding problems they face. The flooding is linked to the anarchic occupation of the area, with no respect for the town planning scheme. The 7th arrondissement

[6] Boutas: Marigots in local Chadian Arabic

is densely populated. "*At one time, the areas occupied by the population were empty, while others were used as water retention basins, still known as marigots. Little by little, the town grew and people began to occupy these pools, which have now become homes, preventing the water from returning to its normal bed.* water, and flooding in particular, remains a real concern that we are still unable to manage, hence the occupation of land for various purposes. The population has grown rapidly in recent decades; according to the second RGPH[7] of 2009, the population of the 7e arrondissement is 223,231. As the years go by, the population continues to multiply. In Chad, the annual growth rate is 3.6%. Referring to the projections made by INSEED[8] on the population of N'Djamena, we take the population figure over 5 years to confirm the strong demographic pressure we are talking about.

Table 3. Population growth in N'Djamena from 2009 to 2019

Year	Population
2009	954200
2010	1005500
2015	1266400
2018	1454671
2019	1521882

Source: French National Institute of Statistics, Economic and Demographic Studies

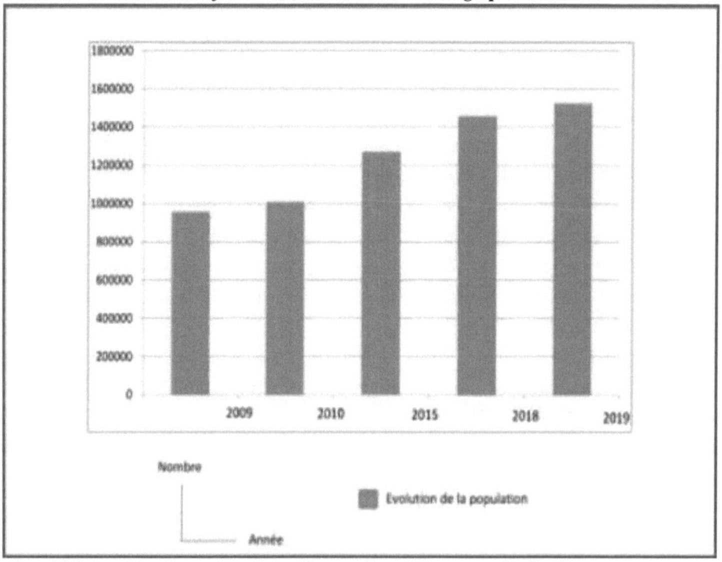

Source: French National Institute of Statistics, Economic and Demographic Studies

Figure 3. Population growth in N'Djamena from 2009 to 2019

Table 3 and the diagram (figure 3) show that the population is changing rapidly, resulting in an uneven distribution of the population over the whole of Chad, particularly in N'Djamena. Faced with this pressure, these inhabitants, in their search for somewhere to live, start hunting for land, resulting in disrespectful occupation and occupying the space intended to

[7] General Population and Housing Census.

[8] Institut National de la Statistique, des Etudes Economiques et Démographiques.

collect rainwater. These areas are occupied by the inhabitants, forcing the marshes and streams to change direction. As a result, the area is almost completely flooded. The houses are flooded and the roads are inaccessible because the occupation has been carried out in an anarchic manner, without taking account of the drains. The exploratory study carried out in these neighbourhoods revealed that there are no drains. Given the high population density, the houses are built to such an extent that the water remains stagnant, making the isolation caused by high demographic pressure the real problem.

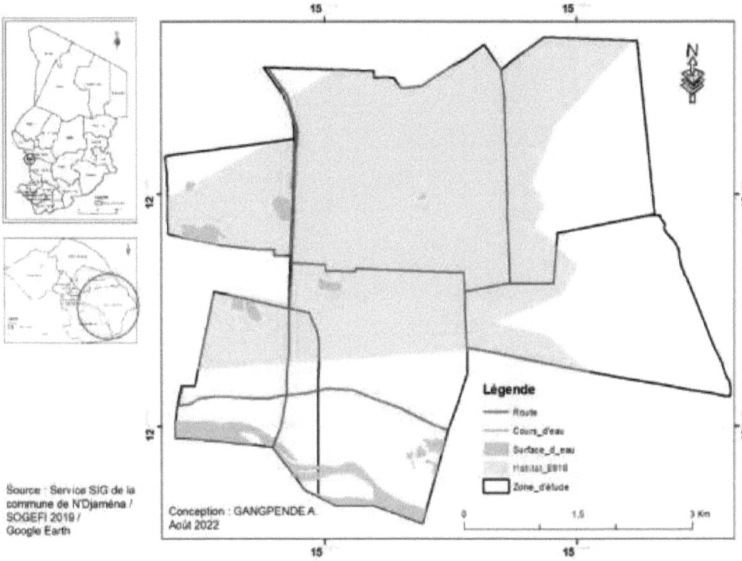

Figure 4. Urban expansion in the study area (Year 2010)

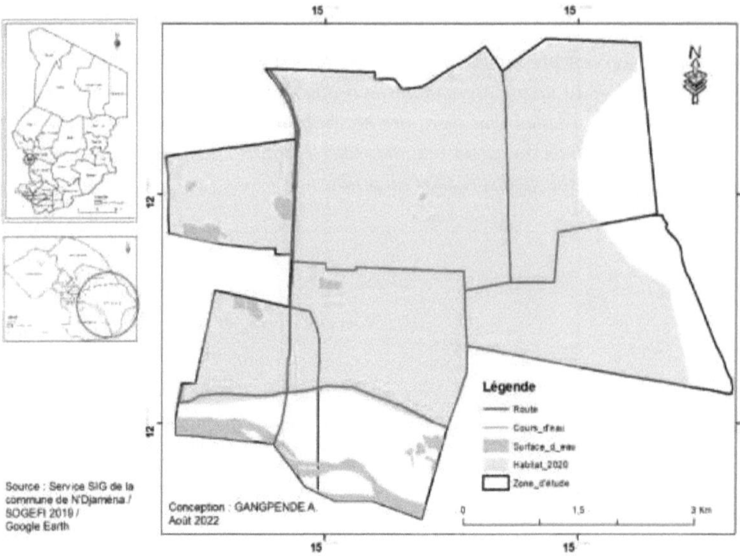

Figure 5. Urban expansion in the study area (Year 2020).

These two maps (Figures 4 and 5) show the number of inhabitants during this decade. We have urban expansion maps for 2010 and 2020. This 10-year interval gives us a better idea of the rate population growth. The more the population grows, the more space is occupied, so that, without realising it, we are occupying the space intended to collect rainwater. The consequence is the population is swallowed up by floodwaters unless there is a simple channelling system to keep them out of harm's way. The migratory aspect of the population is a major factor in the demographic explosion in the 7th arrondissement.

3.2.2. Urban development

Urbanisation leads to a net increase in soil sealing and changes to the natural channels and beds of rainwater as a result of the development work carried out, including the construction of buildings, roads (bitumen) and so on. The consequences of rampant urbanisation are poor water infiltration and a high rate of surface run-off, leading flooding.

It should be noted that the regime of a watercourse depends on the development work carried out in the area. An open space intended for rainwater should not be occupied for housing, because sooner or later, during floods, this water will return to occupy its bed. If it does, it creates a new one, which can be beneficial for the population.

First of all, it should be noted that the story told by the authorities in charge of the city of N'Djamena has never taken enough time to map out the city as it should be, taking into account all the constraints linked to its development.

The built-up areas (homes) outnumber the bare land, yet some of the newly-created outlying districts in the 7th arrondissement'such as Djigo, Siguétté, Kliwiti and Adinio, are virtually empty. Growth is only accentuated in neighbourhoods such as Abena, Atrone, Ambata, Amtoukoui and Boutalbagara, creating a fairly high population density. Houses are being built to occupy the spaces traditionally used for water.

Some of the city's authorities, notably the mayors, have sold or dedicated other areas (marigots) for various purposes (markets, football pitches, etc.). These include the Abena, Atroné, Boutalbagara and Ambata markets. Similarly, the fish market, still known as Tarodona [9][10], is an informal market set up and occupied anarchically by women fish sellers, despite the fact that the authorities have chased them away.

The city of N'Djamena was very quickly invaded by housing. As is often said, houses have sprung up like mushrooms over the last few decades and the water management aspect has always been forgotten.

Figure 6. Aerial view of our study area on Google Earth (31 May 2020)

This aerial view (Figure 6) shows a concentrated built-up area as a result of demographic pressure through construction (roads, buildings, etc.). Our study area covers a surface area of 25.163 km $^{(210)}$. almost entirely occupied by dwellings, the main cause of flooding. In some parts of the area, there are no drains at all, as residents told us during our surveys.

A number of roadworks have been carried out, but it is worth noting that the routes were laid out in such a way as to bypass the town in order to cope with the traffic, which is a reminder of its military character. This is not an idealistic way of improving the town's urban development. Building roads is all very well, but thinking about drainage is even better. How can we understand that a district is not served by any sewers apart from the small drainage networks created by the population? That's not enough, because we need the involvement of the authorities in charge, but we deplore the lack of sewers.

[9] In local Arabic, it means we have been chased away.

[10] Calculated on Qgis

Figure 7. Aerial view of the water retention basins on Google Earth (31 May 2020)

In 2012, the construction company ENCOBAT began building drains and some water retention basins in the 7th arrondissement, but the work has not been completed. Normally, a canal should be built with a slope to drain off rainwater from the neighbourhoods towards a water source (rivers, lakes, etc.), which is not the case for these "gutters". And as the slope is more towards the houses, the water flows into the houses when it rains. Under the agreement between the company and the 7th arrondissement municipality, two water retention basins were to be dug to drain rainwater to the river. But the basins are not even 2 metres deep. So, as soon as the first rains fall, these basins and gutters are already full and the water flows into the houses instead of draining into the lake. Flooding then becomes a real and regular occurrence in our study area.

Coordinates: X: 12.0594 Y: 15.0625

Coordinates: X: 12.0623 Y: 15.0584

Plate 1. Same area during and after the flood

Photo: Aristide GANGPENDE

1st photo (02 April 2020 at 13h03:24) and 2nd photo (21 August 2020 at 15h12 :57)

This plate, consisting of two photos (1 and 2) taken in the Abena district, shows a realistic situation photographed in the same place at two different seasons (flood and deflooding). On this plate, we can see that in the 1st photo the area is free of water during the flood season, during the flood season the area is completely flooded. The pipe (indicated on the photo by the red arrow) built by the company was useless. As the water level is low, as soon as it rains, the canal fills with water, which naturally overflows.

In addition to the issues mentioned above, it is important to highlight the lack of civic-mindedness on the part of the population, which is focused on the issue of insalubrity. People dump their rubbish in the drains. The gutters are clogged with dirt that prevents the normal flow of water under the helpless gaze of the municipal authorities. There is also the problem of a lack of drains in certain neighbourhoods in the area, such as Boutalbagara, Atrone and Ambata, which have no drains at all. Another problem highlighted by the population is the misunderstanding between residents, with neighbourhoods not served by drains draining water to other neighbouring neighbourhoods. Water drained from other neighbourhoods is one of the causes of recurrent flooding in our study area.

Photo 1. Household waste in a drain (unsanitary condition)

Shot: Aristide GANGPENDE on 07 October 2020 at 08h33: 29

Coordinates: X: 12.0599 Y: 15.0590

At the end of chapter 3 on the explanatory factors of flooding in our study area, the main causes of flooding in the Abena, Atrone, Ambata, Amtoukoui and Boutalbagara neighbourhoods were presented. The main causes of flooding in the Abena, Atrone, Ambata, Amtoukoui and Boutalbagara districts were presented, and it was agreed that flooding in our study area is a combination of natural causes (climatic variation and geomorphology) and human causes (demographic pressure taking into account land use and urban development). We started from the observation that climatic variations lead to irregular and sometimes abundant rainfall, not forgetting the morphological aspect of the terrain (slope, basin, etc.) and its impermeable nature, which encourages surface runoff, hence flooding. In the second part, we pointed the finger at demographic pressure, the source of anarchic occupation of the land by construction and various urban development works (roads, buildings, etc.). Similarly, the problems of sewerage in certain districts of our study area and the population's lack of civic-mindedness (insalubrity and sanitation) are also analysed.

CHAPITRE 4. CHARACTERISATION OF WATER-RELATED DISEASES IN THE DISTRICTS OF ABENA, ATRONE, AMBATA, AMTOUKOUI AND BOUTALBAGARA NEIGHBOURHOODS

INTRODUCTION

Water-related diseases are among the most recurrent worldwide, and are often linked to several factors. In recent decades, Chad has often had to deal with issues of flooding, lack of water, hygiene and sanitation. The flooding factors listed in the previous chapter will enable us to answer our question about the health repercussions of flooding in our study area. To do this, we focused on two fundamental points: the main causes of water-related diseases and, secondly, the characteristics of these diseases.

4.1. CLASSIFICATION OF WATER-RELATED DISEASES

There are many diseases that can be transmitted by water and they are classified into several types; however, we will only deal here with those that are water-related. But first of all, it is very important to distinguish between water-borne diseases and water-related diseases. These two terms still cause confusion in some studies.

♦♦♦ ***Water-borne diseases***: in the general sense of the term, water-borne diseases are diseases (and by extension health risks) linked to water quality and access to drinking water. According to Razak (2017), water-related diseases are all illnesses or alterations in a person's state of health directly linked to the use or consumption of poor quality or contaminated water.

♦♦♦ ***Water-related diseases***: according to the World Health Organisation (WHO, 1994), these all diseases indirectly linked to water[11] or diseases caused by the various pathogens found in different water sources.

Table 4. Classification of water-related diseases

Diseases	Germs	Transmission volutes	Vectors	Event	Geographical areas
Paludismo	- *Plasmodium (Falsiparum. malariae, ovale, vivax)*	Blood	- *Anophela (gambiae. funestus, moucheti, arabiensis)*	- Fever - Cephalée - Vomiting	Endemic in the intertropical zone
Typhoid fever	- *Salmonella typhi*	Oral	- Water and contaminated food	- Constipation - Headache - Loss of appetite - Stomach pain - Diarrhoea	In Africa, Latin America, Central and South-East Asia
Gastroenteritis	• *Young Campylobacter* • *Yersina enterocolita* • *Rotavirus, Astrovirus and Fadenovirus* • *Norovirus*	Oral	- Soiled water and food	- Abdominal pain and cramps - Intense constipation - Diarrhoea - Flatulence (excess gas) - Feeling of ballonnement	Across the globe
Bilharzia (Schistosomiasis)	- *Schistosoma (Hemaetobium. mansoni, intercalatum)*	Skin	- Dirty water	- Abdominal pain - Diarrhoea - Genital itching - Papular dermatitis of the feet	Tropical and subtropical zone
Dracunculiasis	- *Guinea worm*	Skin	- Dirty water	- Itching	Tropical and

[11] Quoted by Razak

				- Burning sensation and pain - Fever and rash	subtropical zone
Cholera	- *Vibrión cholerae*	Oral	- Water and contaminated food	- Watery diarrhoea - Intense muscle cramps - Reduced or absent urination - Pressing eyes into Forbite	Sub-Saharan Africa
Dermatosis	- *Staphylococcus sp*	Skin	- Dirty water (direct contact)	- Intense itching	Tropical and subtropical zone

Source adapted from Razak: Pich er al, 2001 and WHO, 2003

Water-related disease pathogens

A pathogen is a factor capable of causing injury or disease (disease process) in animals or plants. A distinction is made between exogenous and endogenous pathogens.[12]

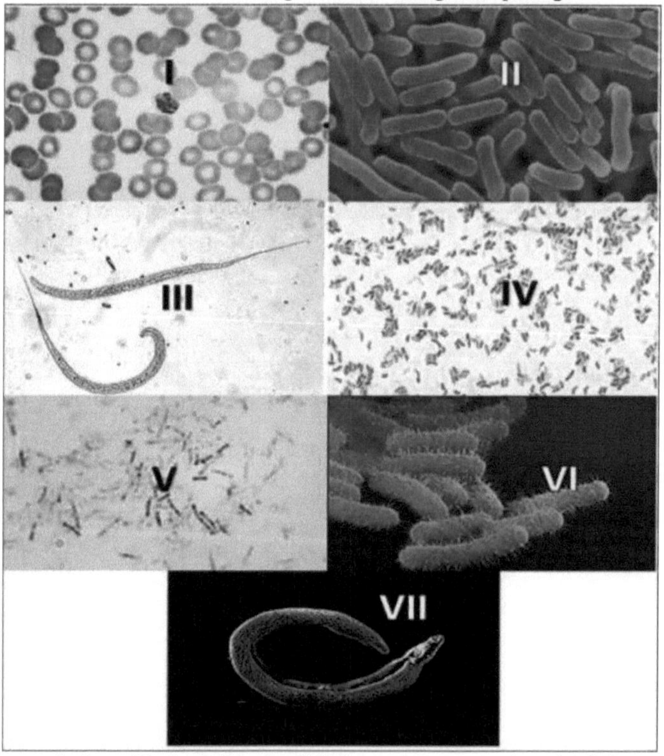

Plate 2. Microscopic views of the bacteria responsible for some recurrent water-related diseases.

Image sources: www.microbes-edu.org and www.vetbact.org

We have a plate (2) made up of 7 images representing the bacteria or agents responsible for water-related diseases seen under the microscope. These bacteria all belong to the Bacillus family, with the exception of Plasmodium, which belongs to the Cocci family. Image (I) shows

[12] https://fr.wikipedia.org/wiki/Agent_pathog%C3%A8ne

***Pasmodium**, image (II) **Salmonella**, image (III) **Guinea worm**, image (IV) **Vibrio cholerae**, image (V) **Dichelobacter Nodosus**, image (VI) **Shigella** and image (VII) **Schistosoma**. Below is a description of these images, based on Google, WHO and www.msdmanuals.com:*

♦ ♦♦ **Plasmodium** is the pathogenic agent of malaria in humans. The parasite is transmitted to humans by the bite of anopheles (a mosquito found mainly in hot regions). In humans and other animals, it first infects hepatocytes (liver cells) and then, during the parasite cycle, erythrocytes (blood cells responsible for transporting oxygen). Other species of plasmodium have been found in various endothermic animals (animals capable of producing internal heat) such as monkeys and birds, but also ectothermic animals (animals living at ambient temperature, the opposite of endothermic animals) such as lizards.

♦ ♦♦ **Salmonella** is a genus of proteobacteria (a group of bacteria) belonging to the Enterobacteriaceae family (found in soil). They measure 0.7 to 1.5 µm[13] in diameter and 2 to 5 µm in length. In humans, they cause diseases such as typhoid fever, paratyphoid fever and salmonellosis, one of the main causes of collective food poisoning.

♦ /- **Guinea worm**, known in Latin as *dracunculus medinensis*, causes a disabling and painful condition which begins with a blister, usually on the leg. At the time of eruption, there may be pruritus, fever, violent pain and burning sensations.

♦ ♦♦ **Vibrio cholerae** is a gram-negative, curved rod-shaped, motile bacterium responsible in humans for cholera, a contagious epidemic disease. This bacterium produces a toxin responsible for the symptoms associated with the disease.

♦ ♦♦ **Dichelobacter nodosiis**, formerly *bacteroides nodosus,* an anaerobe (a species capable of living in an airless environment) and the causative agent of sheep foot rot and dermatitis.

♦ ♦♦ Strictly pathogenic **Shigella**, found exclusively in humans. It was named after the Japanese bacteriologist Kiyoshi Shiga. It is an agent responsible for gastroenteritis.

♦ ♦♦ **Schistosoma**, the pathogenic agent of bilharzia. These parasites, called trematodes infect vessels in the digestive or genitourinary tract.

It should be noted that Shigella and Salmonella are responsible for gastro-enteritis.

4.1.1. Diseases linked to dirty water

These are diseases caused by dirty water contaminated by human, animal or chemical waste. Dirty water-related diseases include malaria, typhoid fever, cholera and gastro-enteritis.

♦♦♦ *Malaria*

Malaria is a term derived from the Latin *paludis*, meaning "*swamp*". An infectious disease caused by the Plasmodium germ, malaria is the most widespread infectious disease in the world and is transmitted by mosquitoes living in stagnant water, dirty water or an unsanitary environment. Also known as malaria, malaria is a parasitic disease that causes fever and digestive problems. It is transmitted to humans mainly through the bite of an infected mosquito (the female of the genus ***Anopheles***), and more rarely through a blood transfusion or mother-to-child transmission during pregnancy.

[13] Micrometer : Device used to measure microscopic dimensions.

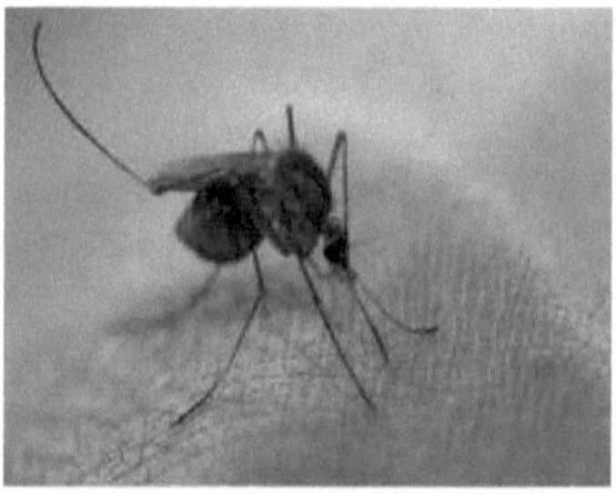

Photo 2. Female Anopheles (photo by Doctissimo.fr)

The photograph (photo 2) shows a female anopheles (malaria vector) resting on human skin. It is easily recognised by its black scales.

In 2015, the WHO recorded 212 million cases of malaria, including 429,000 deaths. Nearly 90% of these cases were in sub-Saharan Africa. Moreover, two-thirds of malaria-related deaths occur in children under the age of 5.

Symptoms of malaria appear 9 to 30 days (depending on the species of Plasmodium) after infection by a mosquito bite. The symptoms include fever, headaches, vomiting, muscle pain and fatigue. If infected by P. falciparum and left untreated for 24 hours, malaria progresses to more severe and often fatal illnesses.[14]

Malaria (mainly P.falciparum[15]) is characterised as serious when a vital organ is affected, either immediately or following a delay in treatment. The complications are varied (respiratory distress, renal failure, coma, convulsions, etc.) and can lead to irreversible after-effects or death.[16]

Humans are diagnosed as being the most affected by this pathogen, through the bite of an infected mosquito (female anopheles). A cycle of reproduction and multiplication takes place in the mosquito, which can then transmit the parasite to other humans. After the bite, in humans, the sporozoites contained in the salivary glands of the anopheles remain in the skin and blood for less than 30 minutes before reaching the liver. The parasites infect the red blood cells, multiply and cause the red blood cells to dissolve. Malaria can persist in the blood and reactivate itself after several years.[17]

♦♦♦ Typhoid fever

Typhoid fever is caused by Salmonella bacteria. People are exposed to the risk of contracting the disease by ingesting water or food contaminated with the faeces or urine of a sick person or chronic carrier. Typhoid fever is a systemic disease of variable intensity. It can range from mild fever to severe systemic disease . The latter is accompanied by abdominal perforation and extra-intestinal infection. People become infected after consuming food or drink that has been handled by an infected person, or through drinking water that has been contaminated with effluent containing the bacteria. Once the bacteria enter a person's body, they multiply and spread from the intestines into the bloodstream).

♦♦♦ Cholera

Cholera is an acute diarrhoeal bacterial infection caused by the Gram-negative bacillus Vibrio cholerae, which produces a toxin responsible for the symptoms associated with the disease.

---

[14] WHO

[15] Plasmodium falciparum

[16] https://www.sante-sur-le-net.com/maladies/maladies-parasitaires/paludisme/

[17] http://www.inrs.fr/publications/bdd/eficatt/fiche.html?refINRS=EFICATT Malaria

Cholera is endemic in the humid tropics of Africa and Asia, and epidemic in some Latin American countries and in dry areas (Sahel). Africa is the continent most affected, with reported cases accounting for more than half of all cases worldwide every year.

Infection occurs through the ingestion contaminated water or food. Most infected people show no symptoms. Signs of the disease appear in 25% of affected patients: liquid diarrhoea in spurts (several litres a day), vomiting and feelings anxiety.

In our study area, case of cholera has been recorded in the last two decades, according to health centre managers.

♦♦♦ ***Gastroenteritis***

Gastroenteritis is one of the main reasons for consulting a doctor, especially a paediatrician. Gastroenteritis often occurs in children under the age of three. are usually infections characterised by acute digestive problems (diarrhoea, abdominal pain, vomiting), accompanied by general signs that are usually discreet. The agents responsible for gastro-enteritis can be bacteria, parasites or viruses: bacteria (Salmonella, Shigella, Campylobacter, E. coli, Yersinia and Vibrio cholerae) and parasites (Giardia, Cryptosporidium, amoebae, etc.).

In adults, the viruses most often responsible for diarrhoea are noroviruses (Caliciviridae), which often cause food- or water-borne epidemics. Astroviruses are also responsible for gastroenteritis in children, the elderly and the immunocompromised. Enteric adenoviruses are responsible for non-seasonal gastro-enteritis. These viruses multiply in the enterocytes of the small intestine. They are transmitted mainly via the digestive tract (faeco-oral), and contamination can be either direct (person-to-person) or indirect (via contaminated water or food). In tropical countries, on the other hand, gastroenteritis occurs throughout the year, and mortality remains high in developing countries, at around 4,000 to 600,000 deaths a year (WHO) .[18]

Another water-related disease is anaemia. The WHO lists anaemia as a water-related disease. In our study area, specifically in Atrone (Atrone Health Centre), two cases were recorded in 2016, leading to the death of these patients. Apart from these cases, no other cases were detected in the years chosen for our study. The main cause is iron deficiency, the most widespread form of nutritional deficiency in the world. In addition to iron deficiency, a number of infections associated with hygiene, sanitation and water quality and management are major contributors to anaemia.

4.1.2. Diseases linked to contact with dirty water

Although there are diseases that do not require constant contact with water to contract, there are others that result from contact with water.

♦♦♦ ***Bilharzia***

Bilharziasis, also known as schistosomiasis, is a disease that affects tropical areas the most. Millions of people throughout the world suffer from this disease. The disease is contracted through contact with dirty water, which carries the parasites into the skin. There are several types of bilharziasis: intestinal and urinary. In our case, we are talking about urinary bilharziasis because it is specific to hot countries such as those in sub-Saharan Africa. The pathogenic agent of this disease is the schistosoma. The parasites penetrate the skin and migrate to the organs where they settle. This can last from 1 to 6 weeks, causing fever, itching and pain. Bilharzia sufferers will find it difficult to urinate, and even when they do, most of

[18] http://www.microbes-edu.Org/etudiant/gastro-ent:erit:es.html

the time their urine is like blood. If left untreated, this disease could lead to sterility.

♦♦♦ ***Dermatosis***

The generic name Dermatosis also known as Dermatitis refers to all affections of the skin and mucous membranes, regardless of their cause. From the ancient Greek *δέρμα / derma* meaning "skin", dermatitis implies a notion of inflammation of the skin. A dermatitis (sometimes called *eczema*) is an inflammation of the superficial layers of the skin characterised by itching, blistering, redness, swelling and often oozing, crusting and scaling. Known causes include dry skin, contact with a particular substance, certain medicines, varicose veins and constant scratching.

Dermatitis can be a brief reaction to a substance, lasting only a few hours or a day or two. Dermatitis corresponds to a reaction of the skin, severe dryness, scratching lesions, an irritating substance or an allergen.[19]

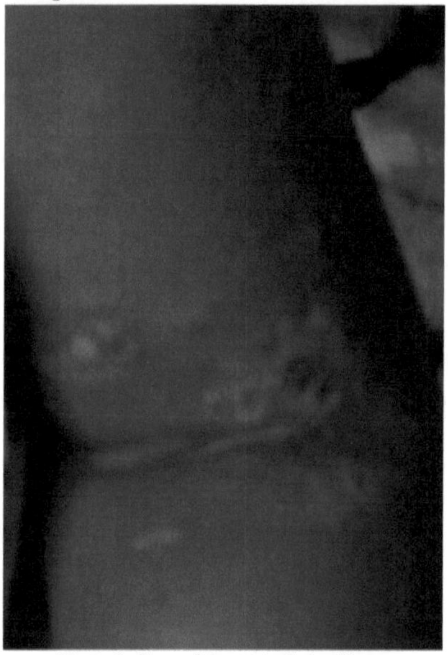

Photo 3. Case of dermatosis in Abena

Photo: Aristide GANGPENDE dated 22 September 2020 at 3:09 pm

Coordinates: X: 15.0626 Y: 12.0593

This photo (3) was filmed in Abena, square 24, in a plot of land about 20 metres from a marsh whose waters spilled into the plot, and shows the affected arm of a 2-year-old girl suffering from dermatitis. The wounds were caused by blisters that had erupted and left sores.

♦♦♦ ***Dracunculiasis (Guinea worm)***

Guinea worm disease, formerly known as medina filariasis, is a parasitic disease caused by the guinea worm or *dracunculus medinensis*. It can be transmitted by ingesting contaminated water. Thread-like, the guinea worm is a parasite that lives in the soil. Detected in

In Chad in 2012, these parasites pierce the skin to lodge in the body. They cause severe pain. People suffering from dracunculiasis find it hard to move around because of the pain. The

[19] https://www.msdmanuals.com/fi7accueil/troubles-cutan%C3%A9s/prurit-et-dermatite/pr%C3%A9sentation-de-la-dermatite

people most at risk are those living near the marshes. There is currently no vaccine against dracunculiasis.

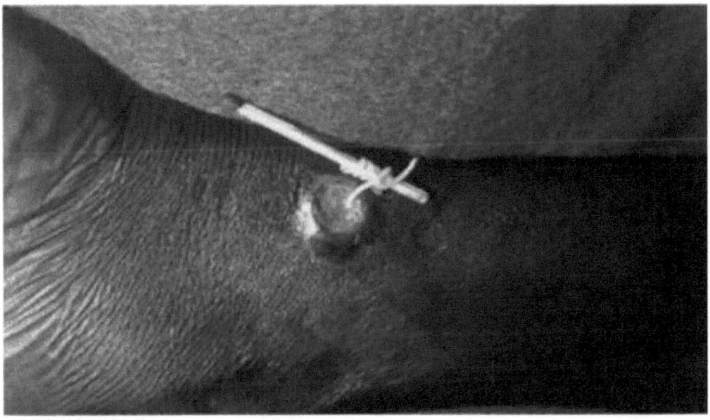

Photo 4. Case of Dracunculiasis

Photo: Médecin Sans Frontière Chad (16 March 2020)

In this photo (4), the foot of a person suffering from guinea worm disease is trying extract the worm from the skin using match stick. The worm can be up to 30 cm long. The worm grows progressively if it is not extracted from the body.

4.2 WATER RELATED DISEASES: Rainwater (flooded and stagnant) and hygiene/sanitation

Water is an essential element for the survival of most species, but in some cases it can also be a risk factor. Water-related diseases all illnesses indirectly linked to water, as well as diseases caused by the various pathogens found in different water sources. Water plays a very important role in human nutrition and in the many activities dictated by living conditions. Quantity, like quality, is one of the indicators to be considered when assessing the level of human satisfaction with water (Toponi, 2013). In our study, both overflowing rainwater and stagnant water have an impact on human health.

4.2.1. Rainwater

Most water-related illnesses in the Abena, Atrone, Amtoukoui, Ambata and Boutalbagara neighbourhoods occur during the rainy season. Water-related diseases can result from direct contact with water or through transmission vectors. According to a WHO study published in 2015, the world records more than 4 billion gastrointestinal infections per year. Infections caused the death of 1.8 million people in 2002. The Third World population is the most affected. In sub-Saharan Africa in particular, the number of cases recorded peaks during the rainy season, highlighting the impact of flooding on people's health.

Rainwater from natural sources is one of the main causes of water-related diseases such as malaria and dermatitis. Water is not a resource like any other: it is fundamental to life. The Abena, Atrone, Amtoukoui, Ambata and Boutalbagara neighbourhoods have received a large amount of rain every year for the past few decades. The rains, which are often more intense or abundant, make it easier for substances and waste to be transported and deposited in the population's water sources.

Periods of rain are considered a blessing, as they give hope to food producers. However, they also accompanied by an upsurge in certain diseases. Residents are regularly exposed to the threat of various water-borne diseases such as malaria, typhoid and dermatitis. According to the consultation and treatment records of the health centres in our study area, children are the most affected and vulnerable group. By often playing in stagnant, dirty water, these children are exposed to microbes that can lead to cases of illness or contamination among them.

In our study area, the consequences of heavy or abundant rainfall are numerous, including the overflowing of water from ponds and gutters, and the contamination of homes whose drinking water sources are close to these ponds and gutters. The result is dampness in these homes, the presence of mosquitoes (Anopheles, the vector of malaria) and contamination of water sources (wells, boreholes, etc.). Four (4) major groups of diseases are linked to the availability and quality of water. These are diseases linked to water contamination, diseases linked to contamination of water by toxic substances and vector-borne diseases. The transmission of malaria by mosquitoes due to stagnant water and gutters in full neighbourhoods illustrates the last category of diseases mentioned.

Photo 5. Stagnant water

Shot: Aristide GANGPENDE on 09 September 2020 at 10h21:12
Coordinates: X: 15.0730 Y: 12.0558

In this photo (5), we can see, in the background, the inhabitants barely standing (broken), and a woman who lives in one of her dwellings going out to fetch water. In the foreground, we see stagnant water in the middle of the concession, caused by the fact that the area is landlocked. This photo shows houses under the influence of rainwater, hemmed in by stagnant water that has no drainage system. Living in this situation exposes local residents to the risk of contamination, as the water contains micro-organisms responsible for the transmission of many water-related diseases.

In our study area, after the rain comes the floods, accompanied by disruption to health and social services. Not only are communication routes (roads) swallowed up by water at this time of year, but homes (concessions, bedrooms, latrines, etc.) are also submerged in water. Not knowing which saint to turn to, these residents are exposed to risks even when they are at home, as no area is spared.

Plate 3. House flooded by water in the Boutalbagara district

Photo credit: Aimée DJIRO dated 07 August 2020 at 07:03 am

This plate (3) is made up of 4 images; the first shows the state of a room engulfed by water in the background and a girl. The second and third images show the bedroom furniture in the water. And finally, in the fourth image, in the background, we see the vegetation made up of a few trees (papaya) and another girl concerned about the state of the house. All the rooms in this concession were filled with water and the inhabitants had to mobilise to get rid of it.

They were forced to work in constant contact with this water, which ended up affecting the health of 6 members of the family (as a result of malaria and typhoïde).

Rainwater is the source of a number of related diseases, the causes of which, although often ignored, are visible. The heavy rains in the 7th arrondissement have led to flooding of roads, schools and other places. Daily activities such as going school, to the market, to work - in short, going about one's business as usual - cannot be carried out without coming into contact with rainwater, which is the source of many diseases.

Coordinates: X: 15.0639 Y: 12.0610 Coordinates: X: 15.0671 Y: 12.0620

Coordinates: X: 15.0606 Y: 12.0611 Coordinates : X: 15.0632

Plate 4. Inaccessibility of roads, schools and workplaces

On this plate (4) made up of four (4) images, the first (1) shows the main road leading to the 7th arrondissement completely flooded, even though it is the main road. The mobility of residents in these conditions exposes them to the risks of dermatitis, malaria and many other diseases. The second (2) and fourth (4) images show pupils going to sit the BEPC/T exams[20] *at the lycée and CEG*[21] *in Abena, while all the school premises were covered in rainwater. The third (3) image shows the state of the Abena health centre. The water from Abena's two water retention basins, made up of faecal matter, rubbish and other waste, is the source of many illnesses suffered by the area's inhabitants.*

Heavy rainfall and constant humidity during the wet season encourage disease.

Added to this is the issue of contamination of water sources (wells, boreholes). It's important to know that overflowing water from these marshes contaminates certain springs close to the marshes. Not to mention the distance between latrines and boreholes/wells. The impermeability of the soil alone does not mean that water is not infiltrating the boreholes.

Coordinates: X: 15.0730 Y: 12.0558

Coordinates: X: 15.0624 Y: 12.0596

Plate 5. Influence of water on water sources (boreholes)

Our plate 5 consists two images filmed by Aristide GANGPENDE, the first (1) on 17 September 2020 in Atrone, showing a borehole near a marsh. According to local residents, when the water overflows, even to access this source of water , you have to step into the water and, when you take the water in a containeryou see a few whitish particles accompanied by a few green particles on the water. Others treat with bleach before eating. If cases of typhoïdes are on the increase in Atrone, it's surely because of this, and the health data don't say otherwise.

The second photo(2) was taken on 20 August 2020 in Abena in a concession in square 24. This water is treated with bleach before consumption. Contamination of this source is the

[20] Brevet d'Etude du Premier Cycle/Chad

[21] Collège d'Enseignement Général located in Abena

cause of several water-related illnesses. This is the case for most of the homes located near marigots or retention basins. Added to this is the problem of hygiene and sanitation, which should be promoted everywhere to put an end to this evil.

4.2.2. Hygiene and sanitation

In 2022, 57% of the world's population (4.6 billion people) had access to safely managed sanitation services. More than 1.5 billion people still lack basic sanitation services, such as private toilets or latrines. Of these, 419 million still defecate in the open, for example in gutters, behind bushes or in bodies of water. In 2020, 44% of domestic wastewater generated worldwide was discharged into the environment without having been treated (1). It is estimated that at least 10% of the world's population consumes food from crops irrigated with wastewater. Poor sanitation is detrimental to human well-being and social and economic development, given its effects in terms of anxiety, risk of sexual assault and reduced educational and professional opportunities. Lack of sanitation is linked to the transmission of diarrhoeal diseases such as cholera and dysentery, as well as typhoid, intestinal worm infections and polio. It exacerbates stunting and contributes to the spread of antimicrobial resistance.[22]

According to some residents, the marigots are the focal point for the activities of local residents. All around, there are children and sometimes older people practising various activities in the marshes surrounding the Abena, Atrone, Ambata, Amtoukoui and Boutalbagara neighbourhoods. Children regularly come here to bathe. Paradoxically, these pools are becoming a destination for all kinds of rubbish from the local population. The waste that ends up there contains a number of dangerous objects, including pieces of glass that can cause cuts leading to infections. In fact, people are more concerned about malaria than other diseases that are just as dangerous as malaria.

To prevent local residents, particularly children, from contracting certain diseases, it is the responsibility of parents to ensure that their children do not play in polluted water. With this in mind, the practice of fishing, known as FOTOK[23] in stagnant water, should be avoided.

When hygiene and sanitation are neglected, residents run the risk of contracting water-related diseases. In fact, rainwater from the Abena, Atrone, Amtoukoui, Ambata and Boutalbagara neighbourhoods is used for a variety of purposes without respecting any rules of hygiene or sanitation. These income-generating activities carried out in these waters are not only beneficial to human life[24] ; they are also vectors of various water-related diseases. Many young people in the Abena neighbourhood now wash their motorbikes, cars and other vehicles. The water in the retention basins is used during the rainy season for washing, and during the cold period at the end of the rains for growing cucumbers. All these ways of using the water from the marshes create numerous risks of illness for the local residents, who are the main users. The

dermatitis and bilharzia are among the main diseases. Children and young people in the Abena, Atrone, Ambata, Amtoukoui and Boutalbagara neighbourhoods all wash their machines.

---

22 https://www.who.int/fr/news-room/fact-sheets/detail/sanitation

23 In local Chadian Arabic, this term refers to a form of artisanal fishing practised in stagnant waters just after the end of the rainy season.

24 Many of the people who wash motorbikes say they do it to survive; they are made up of orphans whose main source of survival is washing motorbikes, or schoolchildren and pupils (and sometimes students) who do it to prepare for the next school year.

Coordinates:X: 15.0671 Y: 12.0620 Coordinates:X: 15.0671 Y: 12.0620

Plate 6. Washing machinery
Source: Journal TchadInfos and Alwihdainfo

Plate 6 is made up of 2 images showing the practice of washing vehicles by young people in the 7th arrondissement. The first image shows a public road (the 10 October axis) and a landscape (trees) in the background. We can also see a few vehicles in traffic and, in the foreground, a few young people washing two-wheeled motorbikes. The second image shows a landscape of buildings in the background, with floodwater in the background. During dry periods, this road is used by machinery but remains inaccessible, except by pirogue, during wet periods marked by flooding. Because of the many activities that young people carry out around and in this wastewater, there is a high risk of contracting related diseases. The water carries the bacteria responsible for a number of diseases.

Fishing for Balboutes[25] by children in the marshes during the rainy seasons is a risk factor for them. For others, these fish, known as Balboutes, are sold in cabarets[26] . The practice of this Balboutes pass exposes practitioners to the risk of disease. According to the consultation and treatment records we consulted at the two health centres in Abena and Atrone, children aged between 10 and 15 are the group most affected by these diseases. As a result, they run not only no risk of contracting the diseases, but also of losing their lives through drowning.

[25] A term in local Chadian Arabic for a species of freshwater fish: the Catfish. This type of fishing is generally practised during the rainy season in marshes. However, even after the rainy season, it can be found in sewage pipes and stagnant water.

[26] Drinking establishment on the outskirts of N'Djamena (7th arrondissement) selling the traditional beer known as BILI-BILI.

Plate 7. Traditional fishing and children playing in stagnant water

Source: Journal TchadInfos and Alwihdainfo.

This plate 7, made up of 2 images, shows the state in which traditional fishing is practised in the stagnant waters of the neighbourhoods in our zone. In the first image, we have a landscape of trees and houses in the background, and in the foreground, children in stagnant water holding the half-cut buckets used to catch the fish. And in the second image, we have in the background the pieces of the jars on the bare ground and, in the foreground, two small children playing in the stagnant water. These two images, the first filmed in Amtoukoui and the second filmed in Ambata by correspondents from TchadInfos and Alwihdainfo with the aim of denouncing the danger faced by these children, perfectly illustrate the risky practices of the local population.

Finally, it is important to highlight population's lack of civic-mindedness when it comes to managing rubbish and household waste. As well as being the cause of flooding[27] , this problem contributes to the stagnation of water.
(pollution).

[27] Water pipes blocked by household waste preventing the circulation of rainwater in our study area.

Plate 8. Unsanitary conditions (before and during the rainy season)

Our plate 8, consisting of 2 images, shows insalubrity before and during the rainy season. The first image shows the rubbish dumped in the water pipes; the second, on the other hand, shows the effect this rubbish has on the quality and circulation of the water. Even in these drains, some children fish for catfish to eat and sell. These pipes, polluted by toilet water and rubbish, are a source of water-related diseases.

People living outside our study area also run health risks (water-related illnesses). But in what way?

It is important to note the migratory or mobile aspect of people in order to better understand the movement of water-related diseases beyond our study area. The movement of people from the Kamnda, Chagoua, Chari mongo and Dembé neighbourhoods, all part of the 7th arrondissement, for various reasons, explains the mobility of the disease. When travelling to the market (Abena, Tarodona, etc.), to school, to health centres or to visit relatives, these residents are forced use roads flooded with contaminated water. Being in direct contact with this water, they run the risk of contracting infectious diseases such as bilharzia, dermatosis, guinea worm, etc.

While visiting families, others spend the night there. We notice that the Abena, Atrone, Ambata, Amtoukoui and Boutalbagara neighbourhoods are the strongholds of mosquitoes, the main vectors of malaria. By being bitten by mosquitoes (anopheles), they run the risk of contracting the disease. They can also contract the disease by drinking contaminated water. It is therefore important to note that population movement (mobility) increases the risk of contracting water-borne diseases in people living outside the study area.

From all of the above, it should be remembered that we started from a simple observation: flooding can damage people's health. This chapter sheds more light on the main causes of water-related illnesses. It focuses on an analysis of the impact of rainwater on the health of local residents, as well as an analysis of the population's lack of awareness of hygiene and sanitation as causes of water-related illnesses. Finally, at , this chapter looks at the agents responsible for the transmission of water-related diseases, and discusses some of the recurrent diseases observed in our study area. To do this, it calls for a knowledge of the pathophysiology of these diseases, which refers to the typology or characteristics water-related diseases.

III PART

PERCEPTION, STATE OF PLAY, STRATEGIES FOR RESILIENCE IN THE FACE OF FLOODING AND WATER-RELATED DISEASES

PART THREE: PERCEPTIONS AND CURRENT SITUATION, STRATEGIES FOR RESILIENCE IN THE FACE OF FLOODING AND WATER-RELATED DISEASES

The second part of our work focuses on the determining factors of hydrological risks and water-related diseases. After analysis, it should be noted that the flooding problems in the Abena, Atrone, Amtoukoui, Ambata and Boutalbagara neighbourhoods are real. Our attention therefore focused on the repercussions that flooding can have on the health of residents. Knowing these different processes, the aim of this third part is to understand the population's perception of hydrological risks and water-related illnesses, and then to propose strategies to adopt in order to effectively resist the illnesses that this generates. This part is structured around two chapters: one dealing with the perception and state of play of water-related diseases over time, and the other analysing the flood and water-related disease resilience strategies initiated by riverside populations.

CHAPTER 5. PERCEPTION AND STATE OF PLAY OF FLOODING AND WATER-RELATED DISEASES IN THE AREA STUDY

Our previous chapter enabled us understand water-related diseases and how they contract. This chapter analyses the state of flooding and the evolution of water-related diseases by season (before, during and after the rains). This chapter analyses the most prevalent water-related diseases in order to provide an overview of their distribution.

5.1. FLOODING AND WATER-RELATED DISEASES WATER

For any research project, it is very important to take stock of the situation. By providing an overview of the situation of the phenomenon under study, it constitutes a field study of the problem to be tackled. In other words, it involves finding out how the population perceives flooding, because most of the water-related illnesses in the areas studied are caused by flooding.

5.1.1. Perception of flooding and water-related diseases by the in the study area

Knowledge of something enables us to avoid the worst; yet the person who lives in ignorance is undoubtedly running towards a danger whose source he obviously does not know. If a person who has been warned of a danger is always exposed to it, the person who lives in ignorance is even more so.

♦♦♦ ***Public perception of flood risk***

Water is distributed unevenly throughout the area; this vital source is known to everyone, as flooding is a current occurrence in our study area. It is therefore perceived by the population of the areas studied as a whole, even if others do not face up to it (in terms of losses).

Table 5. Perception of flood risk

	Workforce	Percentage	Valid percentage	Cumulative percentage
Public perception of flood risk.	100	100,0	100,0	100,0

Source: Field surveys

Table 5 shows the awareness of flooding among the population of the study area. Our field surveys revealed that 100% of the population has at least heard of flooding. This suggests that flooding is a real phenomenon in our study area. Public action must be taken to curb this problem.

Although the whole population knows what flooding is, only some of them are affected by the damage, with both material and human losses.

Table 6. Influence of flooding on the population of the study area

	Workforce	Percentage	Valid percentage	Cumulative percentage
1-Yes	64	64,0	64,0	64,0
Variable 2-No	36	36,0	36,0	100,0
Total	100	100,0	100,0	

Source: Field surveys

Table 6 divides the population into two parts: those who have experienced flooding at least once and those who have never experienced flooding. The table shows that, out of 100% of the population surveyed, 36% had never experienced flooding, compared with 64% who had experienced flooding at least once.

♦♦♦ *Perception of water-related diseases*

Water-related diseases are part of everyday life for the population of N'Djamena in general, and our study area in particular. Although most local residents are still unaware of the risks associated with flooding, some are.

Table 7. Knowledge of water-related diseases

	Workforce	Percentage	Valid percentage	Cumulative percentage
1-Yes	98	98,0	98,0	98,0
Valid 2-No	2	2,0	2,0	100,0
Total	100	100,0	100,0	

Source: Field surveys

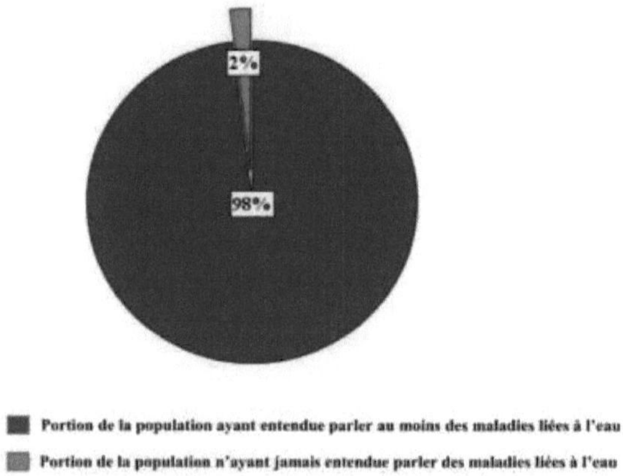

Figure 8. Knowledge of water-related diseases

Table 8 and the diagram (Figure 12) show the perception of water-related diseases by the population in our study area. Of the 100 households surveyed, 98% had heard of water-related diseases at least once, compared with 2% who had never heard of them.

A very important aspect to note is the number of people consulting a health professional in the event of illness, and the number of people practising self-medication. After the 100 households, the percentage suggests that the number of those who self-medicate exceeds those who consult a doctor's advice in the event of illness.

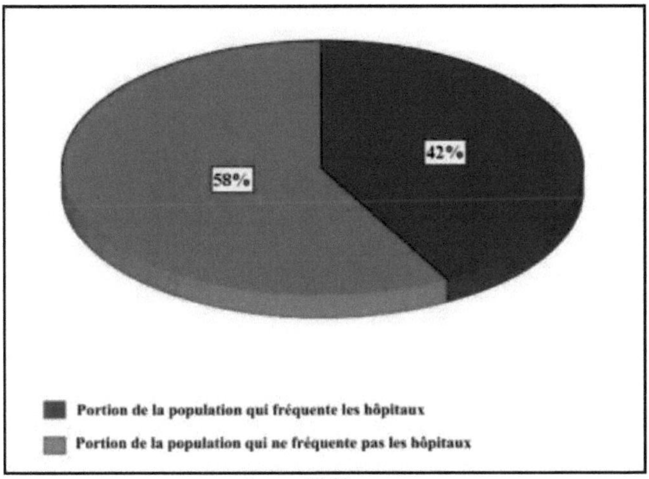

Source: Field surveys

Figure 9. Hospital use by the population in the study area

Figure 9 shows the distribution of the population in our study area in terms self-medication[28] *and hospital consultations. It can be seen that the proportion of the population who do not go to hospital is higher (58%) than that of those who do (42%).*

5.1.2. Trends in water-related diseases by season: dry season and rainy season

Water-related diseases, as their name suggests, are caused by water. But the questions we are asking are whether these diseases are not observed in the dry season? At what time of year are they more prevalent?

Our surveys have shown us the seasonal pattern of water-related diseases. To understand the evolution of a

It is important to study this phenomenon over time. We decided to look at the behaviour of water-related diseases during the rainy seasons (with flooding) and during the period of flood recession. The percentage will determine whether the floods were really the source of water-related diseases, which explains the feasibility of our work. We have tried to find out in which seasons water-related illnesses recur.

[28] According to the WHO, self-medication is the treatment of certain illnesses by patients using authorised medicines that are available without a prescription, safe and effective under the conditions of use indicated.

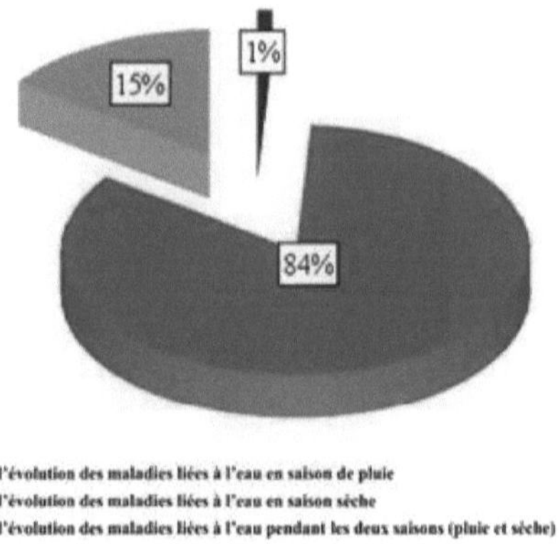

Portion of revolution of water-related diseases in the rainy season
Portion of revolution of water-related diseases in the dry season
Portion of Involution of water-related diseases during the two seasons (rainy and dry)

Source: Field surveys

Figure 10. Seasonal trends in water-related diseases

The diagram (Figure 10) shows the seasonal distribution of water-related illnesses. For every 100 individuals surveyed per household, i.e. a percentage of 100%, the question was asked as to the season in which water-related illnesses recur. In accordance with the variables defined, the analysis shows that 84% of the population stated that water-related illnesses were recurrent during the rainy season, compared with 15% of the population who stated that they were affected during the dry season. The remaining 1% of the population claim to experience water-related illnesses in both the rainy and dry seasons.

For most of the population, these diseases are transmitted through water. So, during the dry season, it is not possible to contract a water-borne disease. For some, even during the dry season, the population is exposed to health risks caused by water. Discussions with health staff at the Abena, Atrone and Ordre de Malte health centres and consultation of consultation data enabled us to note that water-related illnesses are recurrent during the rainy season, even if they sometimes appear during the flooding period.

Table 8. Seasonal trends in Геаи-related diseases for 2016.

	January	**February**	**March**	**April**	**May**	**June**	**July**	**Aoiìt**	**September**	**October**	**November**	**December**	**Total**	**Percentage**
Malaria	40	19	45	25	17	60	35	134	255	50	30	70	780	93,86%
Typhoid	2	9	3	3	0	1	1	1	2	1	4	11	38	4,57%
Bilharzia	0	0	0	0	0	0	0	0	0	0	0	0	0	0,00%
Cholera	0	0	0	0	0	0	0	0	0	0	0	0	0	0,00%
Gastroenteritis	1	0	0	0	1	0	4	0	0	2	2	1	11	1,32%
Dermat dares	1	1	0	0	0	0	0	0	0	0	0	0	2	0,24%
Guinea worm	0	0	0	0	0	0	0	0	0	0	0	0	0	0,00%
													831	**100,00**

		%

Source: Atroné health centre consultation and care records

Table 9. Seasonal trends in Геаи-related diseases for 2017.

	January	February	March	April	May	June	July	August	September	October	November	December	Total	Percentage
Malaria	30	13	40	21	12	51	72	179	148	51	24	43	684	78,44%
Typhoid e	4	1	2	0	11	3	11	41	23	7	3	2	108	12,39%
Bilharzia	0	0	0	0	10	0	0	9	3	0	0	0	22	2,52%
Cholera	0	0	0	0	0	0	0	0	0	0	0	0	0	0,00%
Gastroenteritis	0	0	1	0	0	2	6	2	1	0	0	1	13	1,49%
Dermatosis	1	0	0	0	0	4	9	20	7	1	2	1	45	5,16%
Guinea worm	0	0	0	0	0	0	0	0	0	0	0	0	0	0,00%
													872	**100,00%**

Source: Atroné health centre consultation and care records

Table 10. Seasonal trends in water-related diseases for 2018

	January	February	March	April	May	June	July	Aoiìt	September	October	November	December	Total	Percentage
Malaria	13	17	40	207	69	80	97	230	156	65	35	40	1049	67,37%
Typhoid	1	4	1	3	7	1	10	57	10	5	1	3	103	6,62%
Bilharzia	0	0	0	0	0	0	0	129	0	0	0	0	129	8,29%
Cholera	0	0	0	0	0	0	0	0	0	0	0	0	0	0,00%
Gastroenteritis	1	0	4	6	1	4	9	3	1	7	0	5	41	2,63%
Dermatosis	1	0	11	20	2	10	21	127	24	6	0	13	235	15,09%
Guinea worm	0	0	0	0	0	0	0	0	0	0	0	0	0	0,00%
													1557	**100,00%**

Source: Abena health centre consultation and care records

Table 11. Seasonal trends in water-related diseases for 2019

	January	February	March	April	May	June	July	August	September	October	November	December	Total	Percentage
Malaria	60	35	6	40	43	47	150	138	91	90	100	84	884	81,18%
Typhoid e	10	9	1	5	7	9	2	3	2	12	13	29	102	9,37%
Bilharzia	0	2	0	1	1	4	7	15	3	0	0	0	33	3,03%
Cholera	0	0	0	0	0	0	0	0	0	0	0	0	0	0,00%
Gastroent	0	1	0	0	0	0	0	0	0	0	0	0	1	0,09%

eritis														
Dermatosis	1	2	0	1	1	0	9	50	0	1	0	0	65	5,97%
Guinea worm	0	0	0	0	0	0	0	1	3	0	0	0	4	0,37%
													1089	**100,00%**

Source: Abena health centre consultation and care records

Tables 8, 9, 10 and 11 show the seasonal evolution of water-related diseases over four (4) years, namely 2016, 2017, 2018 and 2019. It can be seen that water-related illnesses are recurrent during the flooding period (May to November). Some cases are also recorded during the flood-recession period (December to April), which suggests that water-related diseases are contractible in both the wet and dry seasons. In all the years selected, we observed a high rate of cases of malaria, which always dominated, while the rates of cholera, guinea worm and bilharzia were almost non-existent.

In August 2018, the health centre recorded a high rate of bilharziasis, with 129 cases. According to the doctors, this was the first time this rate had been recorded. Bilharzia is second only to malaria, with typhoid fever a close second. According to the WHO (2020), Chad recorded 48 cases of dracunculiasis (guinea worm) in 2019. In the Abena health centre, four cases were recorded. Guinea worm disease has almost been eradicated; the condition is rarely fatal, but sufferers are at risk of disability for a given period.

The monthly breakdowns show the evolution of water-related illnesses, and it is also important to know which of these illnesses are recurrent or frequent.

5.2. STATUS OF WATER-RELATED DISEASES IN THE STUDY AREA

5.2.1. Annual distribution of water-related diseases using registers health

Data from the registers of the Abena and Atrone health centres show that water-related diseases affect the inhabitants of our study area and its surroundings. Apart from cholera, which has been almost completely eradicated in recent years by the authorities in charge of the Ministry of Health in partnership with the WHO, the study area is faced with cases of malaria, typhoid, Guinea worm, dermatitis, bilharzia and gastro-enteritis. A time-series analysis using data from four (4) years, 2016, 2017, 2018 and 2019 respectively, has enabled us to study the evolutionary dynamics of the disease over time.

enabled us to study the evolutionary dynamics of water-related diseases in our study area. The study was based more on the health records of the two centres mentioned above, expressed in a table below (table 12). **Table 12. Distribution of water-related diseases from 2016 to 2019**

Diseases	Year							
	2016	**For % of**	**2017**	**For% of**	**2018**	**For % of**	**2019**	**For % of**
Malaria	780	93,86%	684	78,44%	1049	67,37%	884	81,18%
Typhoid	38	4,57%	108	12,39%	103	6,62%	102	9,37%
Bilharzia	0	0,00%	22	2,52%	129	8,29%	33	3,03%
Cholera	0	0,00%	0	0,00%	0	0,00%	0	0,00%
Gastro-Enteritis	11	1,32%	13	1,49%	41	2,63%	1	0,09%
Dermatosis	2	0,24%	45	5,16%	235	15,09%	65	5,97%
Guinea worm	0	0,00%	0	0,00%	0	0,00%	4	0,37%
Total	**831**	**100,00%**	**872**	**100,00%**	**1557**	**100,00%**	**1089**	**100,00%**

Source: Abena and Atrone health centre consultation and care records

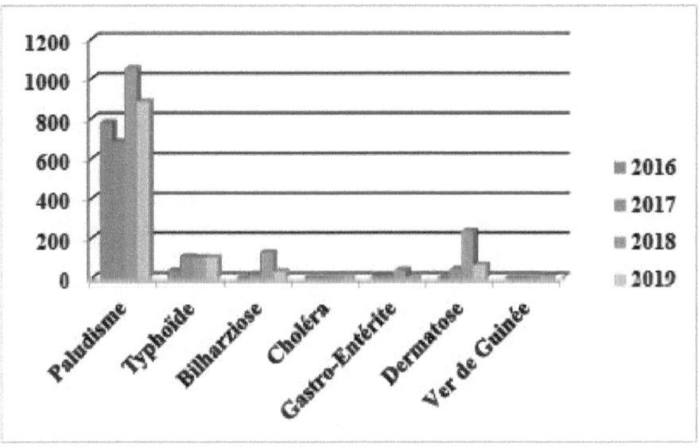

Figure 11. Distribution of water-related diseases from 2016 to 2019.

The data collected shows that the percentages over time are striking. The rate of malaria is the most remarkable, with a rate of 93.86% in 2016 and 81.18% in 2019, followed by typhoid fever and dermatitis, the figures for which barely reach 10%. Other diseases, such as bilharzia, gastroenteritis and dracunculiasis, account for just 4%. Cholera is at the bottom of the scale, with 0 cases recorded during the years studied, broken down as follows (Figure 11) below, showing the dominance of one disease over another others.

5.2.2. Prevalence of water-related diseases

Prevalence is a term used in the health sector to designate the number of cases recorded, including both new and old cases of disease or any other medical event in a given population, either at a specific time or over a given period.

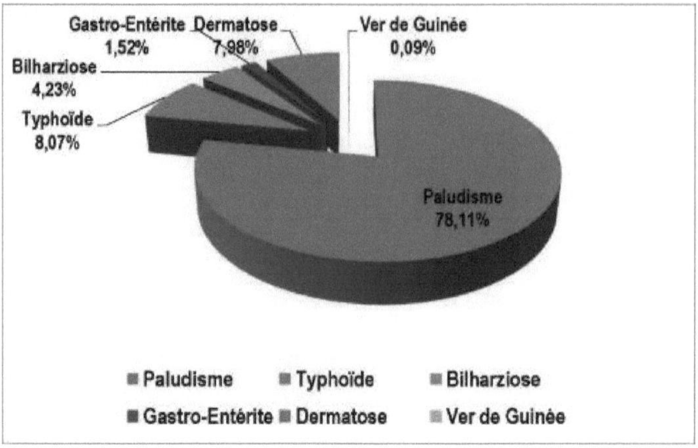

Figure 12. Annual dominance of the diseases studied over each other from 2016 to 2019.

The diagram (Figure 12) shows the dominance of water-related diseases over each other. The diagram shows that the most dominant disease is malaria, with a rate of 74.11%, followed by typhoid fever, dermatosis and bilharzia. Gastroenteritis and guinea worm are almost on the same line, and cholera is absent. Getting closer to the population to get their version of events is an asset for confirming these data collected in the health registers of the health centres. It should already be noted that most of the population self-medicate. This was done in the previous section, but it should be noted that our small survey of the target population was very useful in studying the cases in order to compare the data from the health centres.

The aim of this chapter was to take stock of the situation in order to understand the evolutionary dynamics of water-related diseases. The evolution and dominance of water-related diseases expressed through tables, diagrams, etc. The population is much more familiar with malaria as the main water-related disease. Yet other diseases continue to wreak havoc. The one that is always underestimated is anaemia, which is more dangerous. It caused two deaths in 2016 at the Atrone health centre and one death in 2018 at the Abena health centre. Anemia is malaria that has degenerated as a result of parasites affecting the red blood cells, reducing the level of blood in the human body. In short, the severity of malaria leads to anaemia.

CHAPTER 6. RESILIENCE STRATEGIES IN THE FACE OF THE RESURGENCE OF FLOODING AND WATER-RELATED DISEASES IN THE NEIGHBOURHOODS OF ABENA, ATRONE, AMBATA, AMTOUKOUI AND BOUTALBAGARA NEIGHBOURHOODS

It is true that water is a common asset because it plays a vital role in human survival. It is also of great importance to a country's economy, as it not only keeps people functioning (avoiding dehydration) but also serves a variety of purposes, including agriculture, household chores and others. But it should not be forgotten that its misuse can lead to extreme damage. Already unevenly distributed over the earth in terms of both quality and quantity, rainwater has, more often than not, given rise to a great deal of destruction around the world, and more specifically in our study area. Faced with the damage caused by flooding and the resulting illnesses, the aim of this chapter is to propose a number of possible solutions to complement existing ones, in order to limit the population's vulnerability to hydrological and health risks. The main aim of this study is contribute to the construction of our and to participate in its sustainable and integrated development.

6.1. EXISTING RESILIENCE STRATEGIES TO FLOODING FLOODING AND WATER-RELATED DISEASES IN THE ABENA, ATRONE, AMTOUKOUI, AMBATA AND BOUTALBAGARA

Floods and water-related diseases are the major issues in our study. Although the focus is more on water-related diseases, it was felt useful to touch on both flood and water-related disease resilience strategies.

6.1.1. Existing flood strategies

♦♦♦ ***Contributions from those involved in the fight against flooding***

To avoid any extreme and disastrous consequences, flood management must involve both sides of the problem, namely those affected (the population) and the managers (the authorities in the area). These players must work hand in hand to get the area out of recurrent flooding.

Flooding caused by heavy rainfall has repercussions and poses a risk to the lives of local people and their ecosystems. The inhabitants of our study area are vulnerable every year, and we are well aware that flooding has serious social and economic consequences. The aim of this theme is to propose some possible solutions to reduce flooding and to be vigilant for future floods. This will shed light on the public safety of the area's inhabitants and help the most vulnerable sections of society.

For years, those involved in the fight against flooding have been working to possible solutions. Faced with the damage to their homes, the Chadian authorities and humanitarian organisations take the trouble every year to talk to local residents about the threats. Chad is very often faced with flooding, which once again puts the population in disarray, not to mention the SELS phenomenon, which is the hardest period after the rainy season for certain districts of N'Djamena. The mayor of the municipality, in partnership with the Ministry of Town and Country Planning and ONAT[29] , meets every year to propose solutions to help the inhabitants out of the crisis. However, these solutions remain ineffective. The body directly involved in managing this crisis is the municipality of the 7th arrondissement, more specifically the roads department. By way of illustration, the actions undertaken by ONAT in collaboration with urban planners, geographers, engineers, sociologists and economists to

[29] National Order of Architects of Chad

propose several strategies to get the communes of the 7th, 9(th) and 10th arrondissements out of the cyclical flooding phenomena are part of a perspective of ending the crisis.

Plate 9. Press conference to propose flood control strategies by ONAT and urban stakeholders
Photo Pr Tob-rô (Teacher at UNABA)[30]
As of 26 August 2020

Plate 9, consisting 4 images, shows the press conference organised by platform bringing together ONAT, town planners, geographers, engineers, sociologists and economists with the aim of proposing solutions to flooding. These two organisations note that the recurrence of flooding in recent years, resulting in heavy loss of life, is causing enormous damage to local populations. The lack of responsibilities devolved state institutions undermines the issue of how to deal with the city's sanitation, especially given the emergence of local initiatives to deal with flooding without technical support. At the end of their meeting, they proposed a number of strategies, including increasing the number of pumping stations to water towards the river Chari or the city's peripheral basins, and involving construction companies, technical development partners and volunteers in the cleaning and drainage of gutters to evacuate certain basins and any water retention areas. Identifying areas not affected by the flooding and proposing emergency accommodation in tents or container-based shelters to house the families affected.[31]

In addition to these actions, the committee of the Association des jeunes pour la riposte contre l'inondation (AJRCI[32]), headed by young woman Nétoua Ernestine, has mobilised for citizen action improve roads and assess water drainage.

[30] Adam Barka University, Abéché

[31] Tchadinfos.com/tchad-une-plateforme-propose-des-solutions-pour-sauver-ndjamena-des-inondations

[32] Youth association for the fight against flooding

Photo 6. Young people come together to fight flooding

Photo: Nétoua Ernestine

On 28 August 2020

In this photo (6), we see the young people of the district, concerned about the flooding phenomenon, united as one to deal with it. Criss-crossing the main affected neighbourhoods, they are helping those affected both to evacuate the water and to build dykes to prevent the water from invading their homes.

It emerges that these actions, prompted by the urgency of the situation, remain largely palliative solutions, limited and sometimes archaic in the medium and long term. For more effective and lasting action, it is important that all the stakeholders involved in flood control, from the strategic and decision-making level of the State to the municipalities concerned, get involved to provide sustainable solutions for our study area.

♦♦♦ New flood defences

Several development projects have been carried out to deal with the increase in flooding. These include the construction of retention basins and gutters in the Abena and Atrone neighbourhoods in 2012. These retention basins are basically intended to collect water from other neighbourhoods via the gutters, which should direct the flow in the same direction. Although the decision to evict some of the homes built in the areas designated for rainwater collection has now been implemented, it was a real tug of war between the residents and the government. The only purpose of these new development works is to protect the study area from recurrent flooding.

Plate 10. Roadworks in the Abena district

Photo credit: Journal Tchadlnfos

Plate 10, consisting of 3 images, shows development work underway in the Abena district. In the first image, a landscape of houses and trees can be seen in the background, and in the second, a small road under construction now Taiwanese Avenue). On the right (in the corner) of the photo, in addition to a Caterpillar, in the foreground you can see a gutter being dug (which has barely begun). In the second image, on the other hand, we see a roadwork project being finished, washed away by the rain. In the background is a landscape of trees, and in the second shot is a view of the work being swallowed up by the rainwater. Finally, the third image shows a landscape of trees in the background and a Caterpillar carrying out the work in the foreground.

♦♦♦ ***Educate residents about uncivil behaviour: hygiene and sanitation***

Protect the environment to promote the slogan of a healthy body in a healthy environment. The Chadian artists gathered together to sing to combat insalubrity in the 7[e] arrondissement: Mac Alex, KKJ, Yasmine, Placide and Gladiator.

Photo 7. The Koulina Sawa[33] collective to combat insalubrity
Youtube photo: https://youtu.be/ue9oDKXUdgc

[33] Koulina sawa, which means "all together" in local Chadian Arabic.

This photo 7 was captured in the sound video to insalubrity in the 7th arrondissement. View from a balcón, cloudy landscape in the background, from left to right rapper Kkj, slammer Placide Ayereh, singer Yasmine Abdallah, singer Mac Alex and singer Gladiator. The collective was set up in 2015 to support the local population in the fight against insalubrity. According to a member who told us that due to a lack of adequate resources to support the project, the group has been disbanded. This is part of the lack of existing resilience strategies to which we want to add value.

Every year, the authorities remind residents of hygiene and sanitation issues, in particular to avoid throwing rubbish into the gutters. With this in mind, the government has built large rubbish bins to combat insalubrity. Every year, teams from the local authorities scour the neighbourhoods to remove the rubbish and facilitate the smooth flow of water. Some drains are designed for rainwater only. Others use them to drain water from their toilets. However, this should not be used as a rubbish bin.

These actions are good for fighting and coping with flooding. However, they still do not meet the expectations of residents and those involved in the fight against flooding. We note the neglect of decisions by residents and the lack of follow-up by decision-makers. If you like to play doctor after death, no problem will be solved that way. Words must be accompanied by deeds.

6.1.2. Existing strategies for water-related diseases

♦♦♦ *Contributions from players in the fight against water-related diseases*

Water-related illnesses are a recurring theme in our study area and a major concern. The media are talking about it, and the government is raising people's awareness of the risks. Visit The Ministry of Health, through various programmes, is fighting against water-borne diseases. Thanks to these efforts, cholera has almost been eradicated in N'Djaména, particularly in the Abena, Atrone, Amtoukoui, Ambata and Boutalbagara districts. This calls for a concerted effort to put an end to these recurring water-borne diseases. A number of measures have been taken to combat water-borne diseases, some of them supported by the government and its technical and financial partners. On 13 August 2010, in response to an appeal from the Chadian government, the WHO, through its Representative in Chad, Dr Sai'dou Pathé Barry, provided 4.9 tonnes of medicines and 1,600 impregnated mosquito nets to support the government's response to the floods (WHO, 2013). This includes a batch of 2 medical kits, 2 malaria kits and 1 diarrhoea kit, which provide healthcare for 20,000 people for three (3) months. The kits are made up of medicines, equipment, etc. that will boost the capacity of the health services to deal with the illnesses most frequently encountered in such situations, i.e. malaria, etc.

The management of water-related diseases is being handled by the Ministry of Health, the 7th arrondissement municipality and aid associations for disaster-stricken residents and vulnerable people such as the Fondation Grand Creur (FGD).

A crisis committee has been set up, made up of humanitarian organisations and specialised government departments within the National Committee for Assistance to Flood Victims (CONASI). CONASI reports to the Ministry of the Interior. Although it was set up in 1998, CONASI has just been reactivated following the recurrent floods of the last decade. It is chaired by the Director of Civil Protection. Regional crisis committees have been set up and the humanitarian actors present in these regions have played a leading role in the rapid assessment of needs. These humanitarian actors have already come to the aid of people affected by water-related diseases in times of flooding. The International Federation of the Red Cross has launched an appeal for approximately CFAF 3,026,887 to assist more than 105,585 beneficiaries across Chad. OCHA Chad has created a special page dedicated to this crisis. [33]

All these actions are being carried out with the aim of finding solutions to the upsurge in water-related diseases that have captured our attention. Several actions are being taken by the State to remedy this problem. These include the construction of health centres to serve the inhabitants of the neighbourhoods making up our study area, the campaign to distribute impregnated mosquito nets every year and the programme to combat malaria through the NGO Malaria Consortium.

♦♦♦ ***Raising residents' awareness of hygiene and sanitation issues***

Hygiene and sanitation are ways of combating water-related diseases. When we live in a healthy environment, we flourish because we don't run the risk of contracting diseases from dirty water. The government has opted to build a number of rubbish bins to prevent residents from surrounding themselves with household rubbish, and to avoid throwing rubbish into water pipes. This is because stagnant water can be a vector for disease transmission: mosquitoes. An example of this was the creation of a household waste collection agency with a monthly subscription fee of 2,000 francs, of which 1,000 francs was deducted to cover the running costs of this private company. Public awareness of this practice was stepped up. A better understanding of the consequences of water-related diseases is an asset for the Ministry of Health, which is responsible for public health, to avoid the worst and think of possible strategies or preventive measures. The roads department of the 7th arrondissement travels through the neighbourhoods in our study area with buses, a programme instituted by the Mayor of the city of N'Djamena for each commune: the programme is called N'DJAMENA NADIF[34] . These measures have been taken with the aim of ridding all of N'Djamena's communes of insalubrity and, consequently, of making the entire capital N'Djamena a clean city. The aim is to raise awareness and formally prohibit people from defecating in water sources. Many people use these springs to leave their homes, either by dugout canoe or on foot. In the area of sanitation, faced with collapsed latrines and problems of contaminated water, UNICEF has provided 3 sanitation kits and chlorine to prevent water-borne diseases. UNICEF also has 4,800 water purification boxes (800 in N'Djamena).

- ♦♦ ***Some strategies put in place by local residents***

Faced with the resurgence of disease and unable to stand by and watch, some of the inhabitants of our study area, aware of the dangers of misusing water, have taken the initiative to set up a number of resistance strategies to deal with water-related disease. Our field surveys enabled us to see the attitude of the inhabitants to water-related diseases. Only 20% of the population surveyed had adopted preventive control strategies. We have divided these strategies into two points:

- **Treating water with bleach**: some local residents, aware that their drinking water is contaminated by water from the marigots, use bleach to treat the water before drinking it.

---

[33] *htpp://ochaonline.un.org/chad.*

[34] Meaning clean N'Djamena. Nadif = clean in local Arabic

Bleach eliminates the parasites responsible for transmitting water-borne diseases;

- **Sleeping under mosquito nets**: the mosquitoes in N'Djamena are not indifferent to the inhabitants of our study area. Stagnant water and gutters right next to homes are breeding grounds for mosquitoes. The use of mosquito nets is the best strategy for some people in our study area to avoid water-borne diseases.

In any human endeavour, there are limits, because despite all these decisions, some people are still unaware of the extent of water-related illnesses. It's worth noting that the health centres are too far apart to serve the neighbourhoods; the Atrone health centre alone manages the Atrone, Ambata and Boutalbagara neighbourhoods. Other residents complain about the high cost of hospital treatment. As for the health workers, they complain about the Ministry of Health's failure to follow up on their complaints, because there is a lack of working materials and an insufficient number of staff to work in a chain and cover the needs of the ever-growing population.

6.2. STRATEGIES TO REINFORCE FLOODING AND WATER-RELATED DISEASES IN THE ABENA, ATRONE, AMTOUKOUI, AMBATA AND BOUTALBAGARA

It's true that a number of solutions have been put in place by the government and local residents to deal with the flooding and its after-effects, but the problem is still with us.

6.2.1. Flood reinforcement strategies

♦ ♦♦ ***Development work***

Despite the new developments undertaken by the authorities, the problem has still not been resolved. Given the state of the retention basins, the gutters and the lack of gutters in some of the gutters, as a possible solution we suggest maintaining these basins, which are located in the Abena and Amtoukoui neighbourhoods respectively. The depth of these ponds is negligible (1m maximum), and a little more digging is needed to ensure that the surface can retain a significant amount of rainwater. The ponds in these two districts alone are not enough to alleviate this problem. Some of the ponds in the Atrone, Ambata and Boutalbagara neighbourhoods need to be converted into additional retention basins. There are no drains in the Atrone, Ambata and Boutalbagara neighbourhoods. Apart from a few drains made by local residents to evacuate the water from their neighbourhoods, there are no drains worthy of the name. Resources need to be invested in the construction of gutters to serve these neighbourhoods, taking into account the water level[35] , as the slope plays a very important role in the construction of a canal. The direction flow must also be towards a retention basin or a river. It would therefore be preferable to provide the local authority with cisterns and motor-driven pumps capable extracting the water and pouring it into the river. If large gutters worthy of the name were built along asphalt roads, like the gutters built[36] in the Kabalaye and Ardep Djoumal neighbourhoods, rainwater could circulate more easily and stagnation, the main cause of water-related diseases, could be avoided.

In addition, the population must avoid building in areas designed to collect rainwater (risk zones). They should also respect the water level.

♦ ♦♦ ***Reinforcing hygiene and sanitation measures***

[35] Technique used by engineers for building in areas where water often overflows. It works like this: water is taken into a pipe and the slope is observed, which is reflected in the direction flow. The pipe is graduated in cm.

[36] More than 10 metres deep, running under the tarmac. These neighbourhoods have never experienced flooding since the gutters were built.

It's not a good idea to come and play doctor after death, because more often than not, the gutters are emptied while the water is already invading the population. But how do you get the rubbish out of the gutters? And where do you put it when the areas are inaccessible? It is preferable to carry out work to remove rubbish from the gutters in certain neighbourhoods at least 2 months before the rainy season. Extract the rubbish so that the N'Djamena NADIF company can dispose of it far away, where the rubbish dump is located at the northern exit of N'Djamena. The rubbish in the gutters is not there by itself, but because of the actions of the population. Awareness-raising campaigns and monitoring measures could help to reduce these practices and clean up the gutters.

♦ ♦♦ ***Raising public awareness and understanding***

Emphasis should be placed on raising public awareness of incivism, along with measures formally prohibit the dumping of rubbish in gutters. Raise public awareness through campaigns prevent litter from polluting the environment. Gutters are designed to carry water and are not a dumping ground for rubbish. People need to be made aware of the problem. If rubbish bins have been created, why are the gutters blocked with rubbish? The population needs to question itself. In the media (television, radio, newspapers, etc.), in schools, on plaques in the commune, we need to raise the issue of cleanliness and the inconvenience of rubbish in the gutters. It's true that the population hasn't yet reached a high level of awareness, but we're already thinking about raising people's awareness of the risk culture.

An honest monitoring and control team needs to be set up to check on the progress of the development and redevelopment of the existing basins/channels to prevent funds from being misappropriated. We need to bring together the people working in this field to think about other safe measures to get the area out of this recurring problem.

6.2.2. Reinforcement strategies for water-related diseases

♦♦♦ ***Reinforcing hygiene and sanitation measures***

The problem of sanitation is acute in Chad, creating a major public health challenge. The current target is 35%[37] in line with the MDGs[38] . A healthy body in a healthy environment will be the slogan that the inhabitants must advocate. To achieve this, sanitation must be at the heart of our daily lives, as it prevents many illnesses. Fight against defecation in or near water sources through strict prohibition measures.

♦♦♦ ***Other health measures or strategies***

After a tour of the health centres, we asked for more working materials (beds, stretchers and other working tools) and for the pharmacy, which is still poor, to be supplied with medicines to care for patients. More healthcare staff, because there are more patients than there are staff. To compensate for this limited number of staff, competent health workers need to be recruited and distributed fairly among the health centres in the areas concerned. Similarly, the cost of health care must be reduced to ensure widespread access for the population. For the people living in the study area, not everyone has the financial means to be treated in hospital. The very purpose of up public health centres is to give the poorest sections of the population access to healthcare. However, the reality is that this is still not the case, and the population finds it better to seek treatment from the tchoukous[39] . Even if free care is still a problem, the cost of consultations and treatment must be reduced.

---

[37] Ministry of Water and Sanitation

[38] Sustainable Development Objectives

[39] Pseudo-pharmacists in local Chadian Arabic, which comes from the word tchouktchouk meaning resourcefulness.

Added to this is the problem of the geographical distance of health centres from residential areas. In our study area, made up 5 neighbourhoods, 2 neighbourhoods are without health centres. Don't they need treatment too? The answer is shown in figure (13) below.

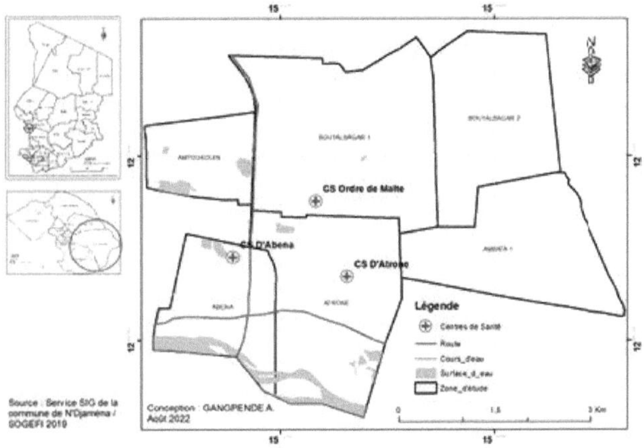

Figure 13. Location of health centres

Figure 13 is a map showing the location of health centres in our study area. On this map, we have 3 health centres for 5 neighbourhoods, which poses a problem. Two health centres need to be built

Two health centres need to be built for the Boutalbagara and Ambata neighbourhoods, as the Atrone health centre alone supplies the three neighbourhoods. is far from the Ambata and Boutalbagara districts, which is why others prefer to self-medicate.

CONCLUSION

In this chapter, after analysing the existing strategies used by city dwellers to deal with water-related diseases, we propose a number of measures or strategies to strengthen these strategies, taking into account possible solutions. So as not to remain indifferent in the face of these problems, the local authorities and residents have proposed a number of possible solutions to control flooding and stem the illnesses. Flooding and water-related diseases must be seen as everyone's business, and their management must be our fight for sustainable, controlled development.

CONCLUSION OF PART THREE

The state of play, prevalence and strategies against flooding and water-related diseases were at the heart of our analysis in these chapters. By the end of this section, it is clear that everyone is aware of flooding and water-related diseases, even if the focus is still on malaria, which is the best-known disease. The survey enabled us to see how the population perceives flooding and water-related diseases. Throughout the study area, the dominance of one disease over another was verified by field surveys and health data from the Abena/Atrone health centres. These health data do not contradict the surveys because they confirm what the target population said. If flooding and water-related diseases are a reality in the area, it is a question for us of proposing strategies with the aim of making a contribution to the construction of this area or this commune of the 7th arrondissement.

GENERAL CONCLUSION

From all of the above, we have discussed a scientific approach that apprehends the factors of floods and water-related diseases and the means of responding to floods and water-related diseases through the strategies proposed by local populations and governmental and non-governmental organisations. Our theme straddles three areas of geography, namely urban hydrology, health geography and behavioural geography.

The main aim of this study is to analyse the problem and find solutions to improve the living conditions of the people living in the area. Starting from an observation made over 5 successive years (2016 and 2020), also belonging to this area, we noticed that our neighbourhood and its surroundings are always in the water. It was this observation that formed the first motive for our exploratory field study.

We asked ourselves a number of questions to find out flooding has any effect on the health of the population. This led us to set ourselves the main objective of showing whether flooding is really the main cause of water-related illnesses. This analysis was based on field surveys of the target population, demographic data, rainfall data and health centre data.

The aim of this study is to test the hypotheses we have set ourselves.

Several factors, both natural (climatic variation and geomorphology) and man-made (demographic pressure and urban expansion), are responsible for flooding in our study area. The combination of these two factors exacerbates the population's vulnerability to flooding.

Flooding in the Abena, Atrone, Ambata, Amtoukoui and Boutalbagara neighbourhoods is having an effect on human health. Water-related illnesses in children account for more than 35% of the top ten causes of consultation in N'Djamena. Malaria is the most widespread disease, followed by typhoid fever and other diseases listed throughout the work.

Water-related diseases continue to wreak havoc in the city of N'Djamena. These diseases are now considered to be a public health problem. Polluted water cause serious illness.

This has been demonstrated by the way floods affect human health. Contraction by contact with contaminated water, contraction by transmission vectors without being in contact with contaminated water. These contraction routes have been demonstrated, as have the pathogens that transmit them and what needs to be done to prevent these diseases.

Finally, it should be noted that the slogan remains the same: "health for all, a healthy body in a healthy environment", and to achieve this, everything must be done to protect the population from floods and disease. This will require a partnership between the population and the authorities in charge, whose only weapon in the fight against flooding will be "controlled and integrated management for the sustainable development of our city".

INTEREST OF THE SUBJECT

The aim of this study is to identify the factors that encourage flooding, water-related diseases and strategies for resilience to these diseases.

From a methodological point of view, the work plan is drawn up, highlighting the documentation, data collection, analysis and interpretation of the results, which constitute, among other things, the approach to be followed in order to produce a scientific work.

From a thematic point of view intended for academic purposes, this work aims to add value in the fields of urban hydrology, behavioural geography, risk geography and health geography. It will provide insight into the determinants of flooding in general and how flooding affects population health.

This work is of applied interest in that it will help the authorities of the city of N'Djamena, specifically the authorities of the commune of the 7th arrondissement, to attach particular importance to the extent to which flooding and water-related diseases are taking place. This work constitutes a database in which the generalities and strategies for preventing water-related diseases will be assigned in order to spare the population from innumerable damage ranging from simple material losses to loss of human life. The results, which we will make available to the public, will raise everyone's awareness so that they can become players rather than mere spectators in this collective struggle.

BIBLIOGRAPHY

Antoine B., Hubert B., Renato S., 2016. *Introduction à la géographie humaine*, Rapport d'enquête, Swiss Tropical Institute, Edition Armand Colin, Paris, 225 p.

Bakari N., 2018. Flooding in the Bebeni district. Master's thesis in Geography, FALSH/University of Ngaoundéré, 140p.

Bétinbaye Y., Goltob M. N., Tchotsoua M., 2015, "Acteurs urbains et gestion de l'eau dans la ville de Sarh (sud du Tchad)", *Syllabus Review, Human and social science series, n°6/2015*, p.69-98.

Courrier Afrique n°43 December/January 2013, p 31 to 35.

Haimes, Y., 1998. Risk management and risk-based decision-making, 8p.

INSEED, 2009. Second General Census of Population and Housing (RGPH2, 2009), 93p.

Jean N.P., 2007. Les maladies liées à l'eau dans le bassin de la volta : état des lieux et perspectives. Institut de Recherche en Sciences de la Santé, Ouagadougou, Burkina Faso 86p.

Kovacs Y., Doussin N., Gaussens M., 2017. Flood risk and cities in developing countries. Technical notes, 111p.

Matembé M.P., 2015. Attitude of determinants of waterborne diseases in children aged 0-5 years. Cas de la zone de santé de Selembao. National Pedagogical University, 69 p.

Mauss M., 2013. *Sociologie et anthropologie,* Paris, les presses universitaires de France, 482p.

Mbayam S. B., 2013. *Morphology, land use and flooding in the city of N'Djamena (Chad)*, Paris, Edilivre, 152 pp.

Ndilbé M. F., 2018. "Stratégies de résilience dans une ville africaine en crise économique: cas du quartier Boutalbagara dans le 7ème arrondissement de N'Djamena", Master's thesis-Research in Sociology, University of Ngaoundéré, 184p.

Nguedjo A. P., 2011. Géomatique et analyse de l'assainissement et des risques sanitaires en milieu urbain : cas du bassin versant du mbanya dans la ville de douala. Professional thesis in Geomatics, GAGER/University of Ngaoundéré, 102p.

OCHA, 2010. Flood Situation Report #1 25 aout 2010, 4p.

OCHA, 2012. Definition de Scenario Préliminaire Inondations N'Djamena - Tchad, 8p.

OCHA, 2012. Humanitarian bulletin Chad September 2012, 6p.

OCHA, 2021. Overview of floods in Chad September 2021, 1p.

OIE, 1995. The world organisation for animal health: health standards and international trade, 6p.

WHO, 2006. World Health Statistics.

WHO, 2008. Primary health care, now more than ever. World Health Report.

WHO, 1994. Urban health crisis. Health for all strategies in the face of rapid urbanization. Geneva.

WHO, 2000. Links between water, sanitation, hygiene and health facts and figures, pp 1-2.

WHO/UNICEF. 2005. Joint Monitoring Programme for Water Supply and Sanitation. Water for Life: Making it Happen. Geneva.

WHO, 2012. Progress on sanitation and water supply.

WHO, 2014. Water-related diseases. https://www.who.int/water sanitationhealth/diseases/en/.

Pardé M., 1959. La crue de la seine en Janvier 1955, 368 p.

Picheral, H., 2001. *Dictionnaire raisonné de la géographie de la santé.* Montpellier. Published by the University Montpellier, 308 p.

Pierre A., Bernard A.G., 2012. Water-related diseases, 7p.
Pierre G., Fernand V., 2006. Dictionnaire de géographie quadrige edition 1979, 250 p.
Situation report https://reports.unocha.org/fr/country/chad/
Razak, A. G. S., 2017. Analysis of the spatial distribution of waterborne diseases in Ngaoundere. In Master's thesis in Geography, FALSH/University of Ngaoundéré-Cameroon. 157p.
Sébastien F., Raymonde S., 2006. Spatialité des enjeux de pouvoir et des inégalités : pour une géographie sociale de la santé, Presse universitaire de Rennes, Edition Armand Colin, p 333-349.
Sylvain A., Réné E.D., Olivier V., 2003. *La géographie cotemporaine*, Paris, Chévalier bleu éditions, 128 p.
Tchotsoua M., Bonvallot J., 1993. Phenomena of erosion and urban management in Yaoundé (Cameroon), 14p.
Toponi, 2013. Contribution à l'étude bactériologique de l'eau de boisson vendue en sachet au marché central de la ville de Kisangani, Mémoire inédit, Fac.de Médecine/UNIKIS, 32p.

Webography

■& Water and health: http://www.eau-seine-normandie.fr/ consulted on 21 April at 9pm.

■ & Hydrography : https://www.universalis.fr/encyclopedie/hydrographie/ Accessed on 21 November 2019 at 16:30.

■ & http://www.cosmovisions.com/Afrique-Hydrographie.htm consulted on 27 December 2019 at 12 noon.

■ www.microbes-edu.org and www.vetbact.org consulted on 22 November 2019.

■ & https://fr.wikipedia.org/wiki/Agent_pathog%C3%A8ne consulted on 17 August 2020 at 1pm.

■ & https://www.sante-sur-le-net.com/maladies/maladies-parasites/malaria/). Accessed on 16 July 2020 at 3pm.

Φhttp://www.inrs.fr/publications/bdd/eficatt/fiche.html?refINRS = EFICATT_Paludism. Accessed on 17 August 2020 at 13:11.

■ & http://www.microbes-edu.org/etudiant/gastro-enterites.html. Accessed on 17 August 2020 at 2pm.

■ & https://www.msdmanuals.com/fr/accueil/troubles-cutan%C3%A9s/pruritus-and-dermatitis/pr%C3%A9sentation-of-dermatitis. Accessed on 17 August 2020 at 15h/

■ & Tchadinfos.com/tchad-une-plateforme-propose-des-solutions-pour-sauver-ndjamena-des-inondations. Accessed on 23 September 2020 at 9am.

■ & htpp://ochaonline.un.org/chad. Accessed on 23 September 9.07am.

■ & Collectif Koulina sawa on YouTube : https://youtu.be/ue9oDKXUdgc. Accessed on 17 December 2020 at 5pm.

■ & SRTM images: www.Earthexplorer.usgs.gov . Accessed on 17 December 2019 at 08:24.

- & Manual of water-related diseases: www.msdmanuals.com . Accessed on 21 November 2020 at 12 noon.
- & https://www.who.int/fr/news-room/fact-sheets/detail/sanitation. Accessed on 22

Flooding and Γeaи-related diseases are major challenges for many urban areas in Africa, particularly in Chad. The 7th arrondissement of the city of N'Djaména, the capital of Chad, faces these problems recurrently, jeopardising the health and well-being of its inhabitants. This study aims to explore the interactions between flooding and Γeaи-related diseases in this specific locality, examining how these phenomena influence each other and analysing the responses and strategies adopted by the local population to cope with these challenges. At the heart of this study are the health implications and socio-economic consequences of flooding and waterborne diseases in the 7th arrondissement of N'Djaména. Indeed, the occurrence of floods can lead to the contamination of Γeaи drinking water, the emergence of diseases such as malaria and diarrhoea, and the deterioration of the living conditions of the inhabitants. Understanding these complex dynamics is essential for developing effective prevention and management strategies for these health risks.

GANGPENDE Aristide was born on 31 August 1994 in N'Djaména, Chad. He holds a Master's degree in Health Geography from the University of Ngaoundéré (Cameroon), and is currently working on a PhD thesis at the University of N'Djaména (Chad), where he is conducting research into the impact of territorial factors on the distribution of typhoid fever in the city of N'Djaména in Chad.

MIX
Papier aus verantwortungsvollen Quellen
Paper from responsible sources
FSC® C105338

Printed by Books on Demand GmbH, Norderstedt / Germany